Innovation in MIMO Technologies, Systems, and Antennas

Edited by Albert Sabban

Published in London, United Kingdom

Innovation in MIMO Technologies, Systems, and Antennas
http://dx.doi.org/10.5772/intechopen.1003546
Edited by Albert Sabban

Contributors
Ahmed El Yousfi, Albert Sabban, Bhawna Kalra, Charles J. Baker, Daniel Segovia Vargas, Daphney Joann, Guisheng Liao, Jaiverdhan, Kerlos Atia Abdalmalak, M.M. Sharma, Steven J. Simske, Vayanaperumal Rajamani, Xin Yuan, Yaxing Yue, Yi Sun, Zhiguo Shi

First published in London, United Kingdom, 2025 by IntechOpen
IntechOpen is the global imprint of INTECHOPEN LIMITED, registered in England and Wales, registration number: 11086078, 167-169 Great Portland Street, London, W1W 5PF, United Kingdom

For EU product safety concerns: IN TECH d.o.o., Prolaz Marije Krucifikse Kozulić 3, 51000 Rijeka, Croatia, info@intechopen.com or visit our website at intechopen.com.

British Library Cataloguing-in-Publication Data
A catalogue record for this book is available from the British Library

Innovation in MIMO Technologies, Systems, and Antennas
Edited by Albert Sabban
p. cm.
Print ISBN 978-0-85014-793-3
Online ISBN 978-0-85014-792-6
eBook (PDF) ISBN 978-0-85014-794-0

If disposing of this product, please recycle the paper responsibly.

Meet the editor

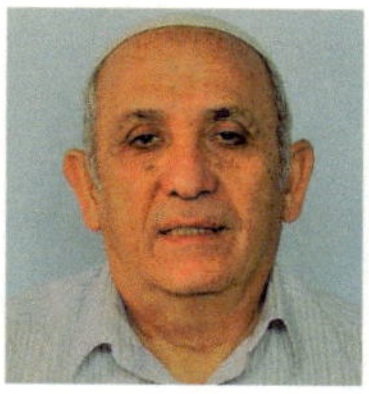

Dr. Albert Sabban holds a Ph.D. in Electrical and Computer Engineering from the Faculty of Electrical and Computer Engineering, University of Colorado at Boulder, USA (1991), and an MBA from the Faculty of Management, Haifa University, Israel (2005). He holds a BSc and MSc (Magna Cum Laude) from Tel Aviv University in Israel. He is a senior lecturer and researcher at electrical and computer engineering colleges in Israel. From 2007 to 2010, Dr. Sabban was a radio frequency and antenna specialist at biomedical hi-tech companies, where he designed wearable compact systems for medical systems. In 1976, he joined Rafael Advanced Defense Systems Ltd., Israel, where he worked as a senior researcher, group leader, and project leader in the electromagnetics department until 2007. He has published more than 100 research papers and holds several patents. He has written and edited over ten books and book chapters on wearable systems, antennas, and green electronics technologies for communication and medical systems. He is the author of a book on electromagnetics and microwave theory for graduate students and a book on wideband microwave technologies for communication and medical applications.

Contents

Preface

MIMO antennas are an antenna technology for wireless communications systems. In MIMO antenna networks, multiple radiating elements are used to transmit and receive signals. The antenna RF signals are combined at each port of the wireless communication system to optimize the communication system specifications, minimize system errors, and improve the system's electrical performance by enabling data to be processed in real-time over many signal RF channels. MIMO antennas and technology are employed for cellular 4G and 5G generation communication technologies and Wi-Fi systems in various applications. MIMO antennas can be used in WLANs, wireless local area networks, law enforcement communication, broadcast TV production, and civil and government applications. MIMO has become an essential element of wireless communication standards such as IEEE 802.11n Wi-Fi 4, IEEE 802.11ac Wi-Fi 4, Long Term Evolution LTE, and more. MIMO networks are techniques for sending and receiving more than one data signal simultaneously over the same RF link by exploiting the difference in signal propagation between different antennas.

MIMO technology increases the data capacity of wireless communication systems by using multiple transmitting and receiving antennas. In MIMO networks, each signal reaches the receiving antenna through a different path, resulting in more reliable data. MIMO technology improves the performance of wireless communication systems.

The book presents several major topics on MIMO antenna arrays and networks. The book presents MIMO antenna systems technologies, massive MIMO communication, cellular networks, and MIMO antenna arrays.

We hope this book will be a valuable resource for researchers, engineers, and project managers involved in developing and manufacturing MIMO arrays and communication systems. The information presented aims to inspire further innovation and application in this field.

Acknowledgments are extended to all the authors who contributed to this volume.

Dr. Albert Sabban
Engineering Department,
Braude College,
Karmiel, Israel

Section 1

Introduction

Chapter 1

Introductory Chapter: MIMO Antennas and Systems

Albert Sabban

1. Introduction

The book will discuss and present new developments and innovation in several major topics in MIMO antennas and systems: MIMO means, multiple input, multiple output antenna systems, see Refs. [1–5]. MIMO is an RF and antennas technology for wireless communications. In MIMO networks, multiple antennas are used as transmitting and the receiving elements. The antennas RF signals are combined at each port of the communication system to minimize system errors, optimize the communication system specifications, and improve the electrical performance of the system by enabling data to be processed in real time over many signal RF channels. MIMO antennas and technology are employed for Wi-Fi systems and cellular 4G and 5G generation communication technologies in a wide range of applications, broadcast TV production, law enforcement communication, and government applications. MIMO can be employed in wireless local area networks (WLANs).

MIMO antennas technology creates multiple versions of the same signal that provides more opportunities for the information to reach the receiving radiating elements without being attenuated and affected by fading and multipath effects, and this increases the signal-to-noise ratio and error rate and improves the system electrical performance.

The wide growth in the development of wireless communications system and devices that require a wideband frequency range, which makes the fifth-generation 5G technology very attractive to increase the capacity, speed, energy efficiency, and reliability of the communication systems, see Refs. [6–11]. Due to the 5G communication system specifications, the needs for wide-bandwidth, light weight, low-profile, and compact size antennas with well-defined radiation specifications and good electrical characteristics are crucial, see Refs. [6–11].

2. MIMO antennas

The continuous growth of the wireless communication industry increased the need to efficient wideband, low-profile, and compact antennas. Moreover, the environmental disaster, climate changes, air pollution, and temperature raise increased the demand to develop green antennas and communication systems. In this introduction chapter, several MIMO antenna radiating elements will be presented. These radiating elements may be used a radiating elements in MIMO antenna networks. Design consideration of development and manufacturing of green MIMO antenna systems will be discussed.

2.1 Printed MIMO antennas

Microstrip patches are compact, flexible, and cheap. Moreover, they have a low profile, light weight, small volume, and low production cost. Printed antennas can be part of MIMO arrays. Microstrip patches are widely discussed in research publications, journals, books, and papers in the last twenty years, see Refs. [6–11]. The popular type of printed antennas is microstrip antennas. However, PIFA, slot, and dipole printed are widely used in communication systems. Printed antennas may be employed in MIMO antenna networks, communication links, seekers, and medical systems.

2.1.1 Applications of printed antennas

- MIMO antennas
- Medical devices
- 5G communication devices
- Medical devices
- Monitoring devices and GPS
- Intelligence and military devices

2.2 Printed patches for MIMO antenna systems

Microstrip patches are etched on a thin dielectric substrate with low dielectric losses. The microstrip patch structure and fields are presented in **Figure 1**. The substrate dielectric constant is εr and the thickness H is less than 0.1λ. Microstrip antennas are widely discussed in research publications and books, [6–11]. Wearable patches can be assembled on human body and in wearable devices. Printed patches are used in wearable smart devices.

2.2.1 Advantages of microstrip patch antennas

- Light weight, low volume, low cost to produce
- Conformal structures are easy to develop.
- Easy to develop large uniform arrays.

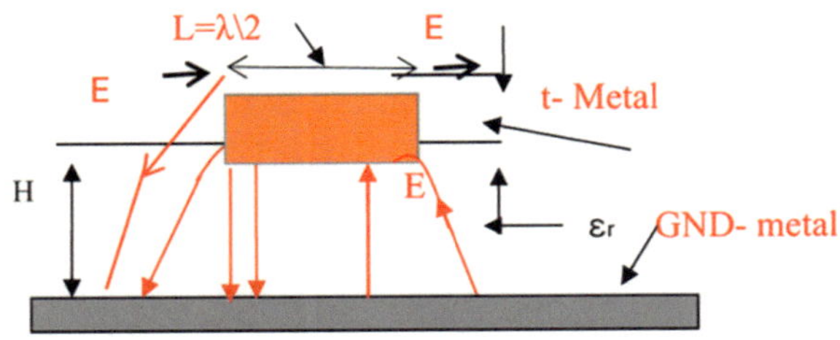

Figure 1.
Microstrip antenna structure and cross section.

2.2.2 Disadvantages of regular patches antennas

- Narrow bandwidth, around 0.5 to 5%. Wider bandwidth can be achieved by using multilayer structures.

- Microstrip antenna can transmit low power rates.

The electromagnetic field along the radiating edges is presented in **Figure 2**. The antenna impedance is proportional to the ratio of E to H fields. Microstrip antennas may be fed by a coaxial cable, microstrip line, or a coaxial probe. By varying the position of the patch feed line between the patch center and the patch edge, we may tune the patch impedance, even to 50 Ω. Microstrip antenna shape may be rectangular, square, triangle, circle, ring, or any arbitrary shape as shown in **Figure 3**. The gain of patch antennas is between 0dBi and 7dBi. The microstrip antenna gain is the function of the antenna dimensions and configuration. We may increase printed antenna gain by using microstrip antenna array configuration. Microstrip antenna feed networks are presented in **Figure 4**. **Figure 4a** presents a parallel feed network.

A square patch antenna dimension W is given by Eq. (1).

$$W = \frac{c}{2f\sqrt{\epsilon_{eff}}} \tag{1}$$

The antenna bandwidth is given in Eq. (2).

$$BW = \frac{H}{\sqrt{\epsilon_{eff}}}. \tag{2}$$

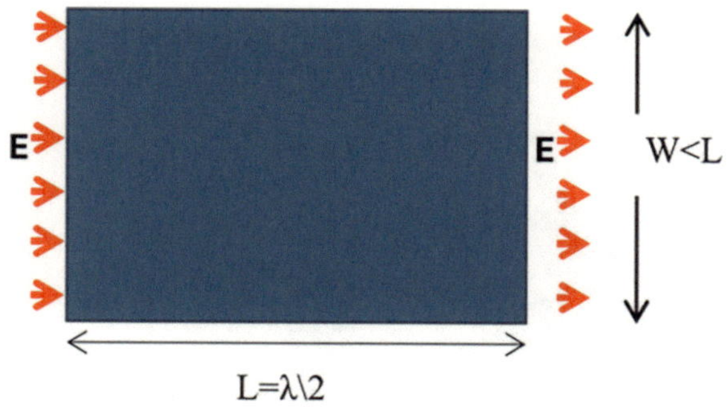

Figure 2.
Rectangular patch antenna structure and fields.

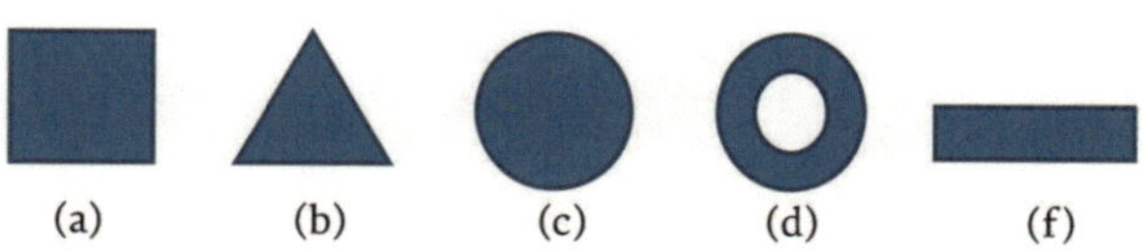

Figure 3.
Microstrip antenna types and shapes (a) Square, (b) Triangle, (c) Circle, (d) Ring, and (e) Dipole.

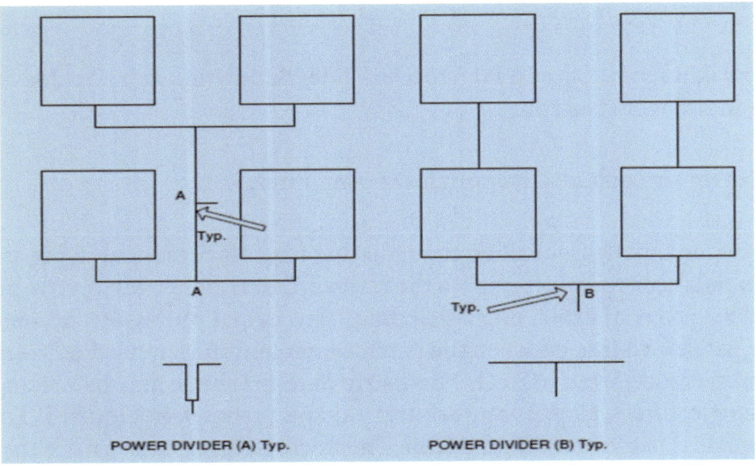

Figure 4.
Microstrip antenna arrays. (a) Parallel feed network. (b) Parallel series feed network.

Figure 4b presents a parallel series feed network. These arrays can be used as MMIO arrays.

2.3 Transmission line model of patch antennas

A good physical understanding of patch microstrip antennas is given by the Transmission Line Model, TLM, as shown in **Figure 5**. The patch antenna is represented as two slots connected by a transmission line. The TLM model is not accurate. The electric field along and underneath the patch depends on the z coordinate, see Eq. (3).

$$E_x \cos\left(\frac{\pi z}{L_{eff}}\right) \tag{3}$$

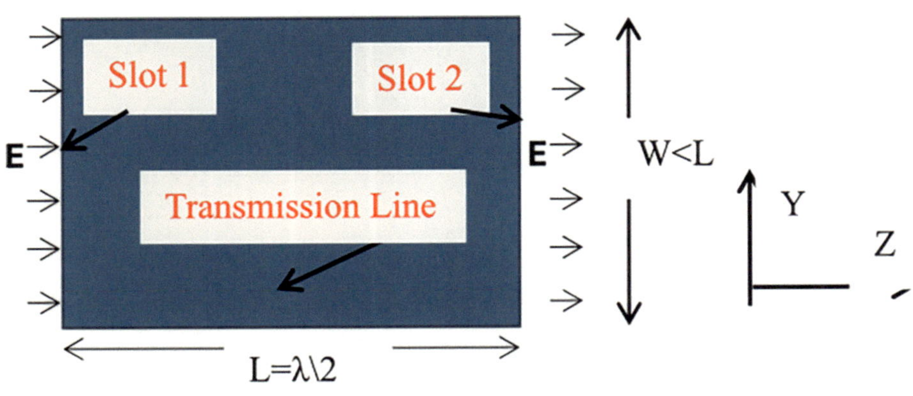

Figure 5.
Transmission line model of patch antenna.

3. Two-layer wearable stacked microstrip antennas

Two-layer microstrip antennas may be the best antenna choice for wideband wearable systems and MIMO arrays. Two-layer microstrip antennas was presented first in Refs. [6–11]. By developing a multilayer patch antenna, we can achieve a wideband microstrip antenna.

On the first dielectric substrate, the antenna feed lines, matching network, and a resonator are etched. In the second dielectric substrate, the radiating element is etched. The resonator electromagnetic fields are coupled to the radiating patch. The distance between the layers can be varied to get the best bandwidth and efficiency. The patch shape can be a circle, ring, square, rectangular, triangle, or any arbitrary shape.

A circular polarized double layer square patch was developed at 2.2GHz. On the first dielectric substrate, with relative dielectric constant of 2.5, the antenna feed lines, matching network, and the resonator are etched. The substrate thickness is 1.6 mm. The patch dimensions are W = L = 4.5 cm. On the second dielectric substrate, with relative dielectric constant of 2.2 and 1.6 mm thickness, the radiating element is etched. The radiating element dimensions are W = L = 4.8 cm. A 3 dB 90° branch coupler is connected to patch feedlines, as shown in **Figure 6** to get circular polarization. The stacked patches bandwidth is around 8–12% for VSWR better than 3:1. The computed and measured stacked patch beam width is 72. The measured stacked patch gain is around 7dBi. This stacked patch may be employed in wearable devices and MIMO arrays. Measured electrical performance of stacked wearable antennas is presented in **Table 1**. The bandwidth of stacked patches is around 8–17% for VSWR better than 3:1. In **Figure 7**, a stacked patch is presented. The resonator feed lines are printed on FR4 substrate with dielectric constant of 4 and 0.16 cm thick. The radiating patch is etched on RT-DUROID 5880 substrate with dielectric constant of 2.2 and 0.16 cm thick. The electrical performance of antennas presented in this chapter was computed and optimized by employing electromagnetic software [12]. The dimensions of the microstrip stacked patch antenna shown in **Figure 8** are 33x20x3.2 mm. The computed S11 parameters are presented in **Figure 9**. Radiation pattern of the microstrip stacked patch is shown in **Figure 10**. The antenna bandwidth is around 5% for VSWR better than 2.5:1.

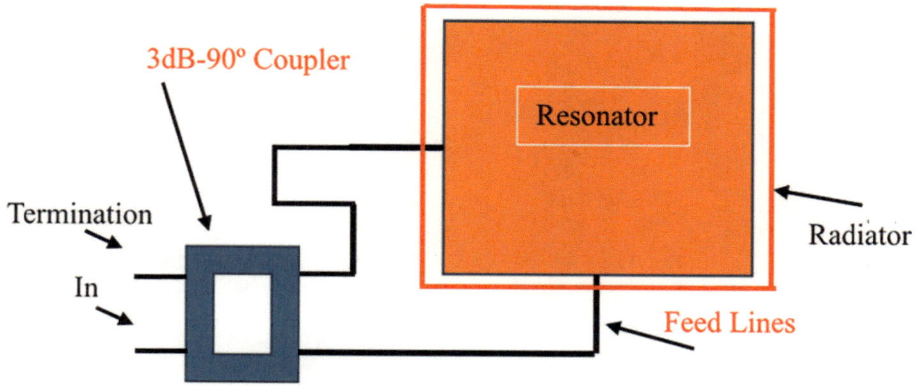

Figure 6.
Circular polarized stacked patch antenna.

Patch shape	F (MHz)	Bandwidth %	Beam-width°°°°	Gain dBi	Side-lobe dB	Polarization
Square	2200	9–11	70–74	7.5–7.7	−21	Circular
Circular patch	2200	13–16	70–74	7.7–7.9	−23	Linear
Annular disc	2200	9–12	76–78	6.5–6.8	−15	Linear
Rectangular	2000	8–11	70–74	7.5–7.6	−24	Linear
Circular patch	2400	8–11	70–74	7–7.2	−21	Linear
Circular patch	2400	8–12	70–74	7.5–7.6	−21	Circular

Table 1.
Measured results of double-layered patch antennas.

Figure 7.
A microstrip-stacked patch antenna.

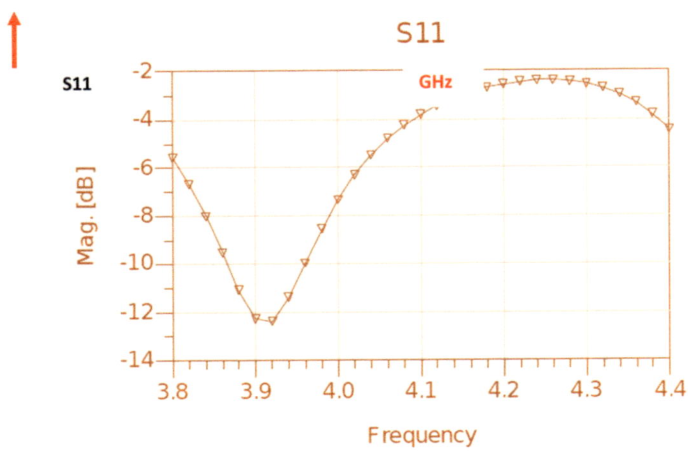

Figure 8.
Computed S11 of the microstrip-stacked patch.

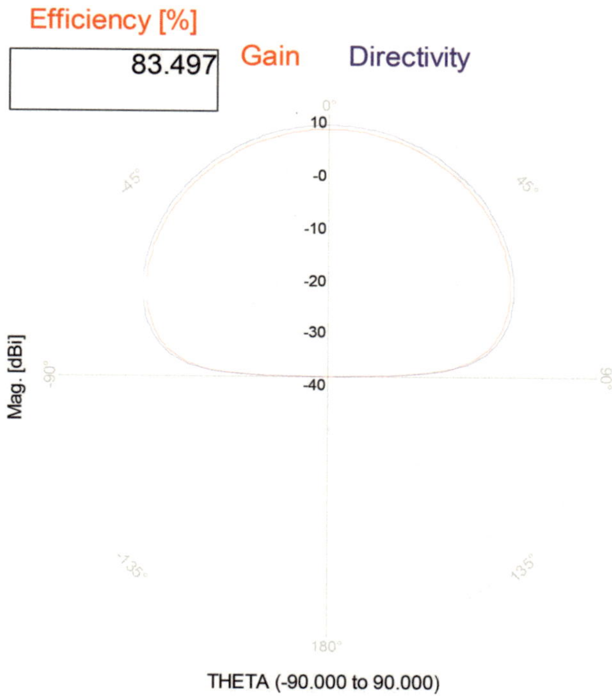

Figure 9.
Radiation pattern of the microstrip-stacked patch.

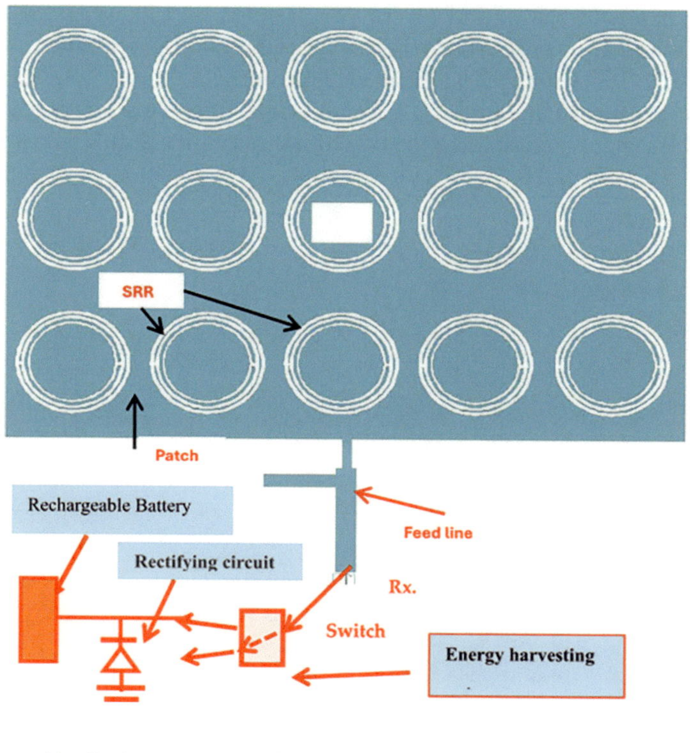

Figure 10.
Microstrip antenna with split ring resonators and an energy-harvesting unit.

For 8-mm air spacing between the layers the antenna bandwidth was improved to 12% for VSWR better than 3.0:1 and the antenna gain is around 7.4dBi.

4. Compact meta materials antennas for MIMO arrays, wireless communication, and medical applications

4.1 Introduction

Printed compact antennas and sensors are widely used in wearable communications, IoT, and monitoring healthcare systems as presented in Refs. [6–11, 13–24]. Materials with periodic artificial structures are known as metamaterials. The artificial material structure and the periodic resonant elements specify the electrical and mechanicals performance of the metamaterial. In Refs. [6–11, 25, 26] CSRRs, periodic circular split ring resonators, are used to produce metamaterials with a dielectric constant value and permeability. Metamaterial antennas may be used as green radiating elements in MIMO arrays, communication systems and wearable medical devices.

4.2 Green microstrip antenna with, CSSR, circular Split ring resonator for MIMO arrays and communication systems

A single-layer metamaterial microstrip antenna with 15 Circular Split Ring Resonators, CSSR, has been designed by employing full wave electromagnetic software [12]. The antenna is printed dielectric substrate with dielectric constant of 2.2 and 1.6 mm thick. The size of the patch antenna presented in **Figure 10** is 3.6x2x0.16 cm. The metamaterial patch bandwidth is around 6% for VSWR better than 2:1 and is around 10% for VSWR better than 3:1. The patch beam width is around 74°. The metamaterial patch gain is around 7.4dBi to 7.6dBi with 95% efficiency. The patch directivity is 8. The measured S11 parameters are shown in **Figure 11**. The gain and directivity of the patch antenna without CSSR is lower by 2.5 dB than the patch antenna with CSSR. Theory of CSSR was presented in detail in Refs. [6–11, 25, 26]. As shown in **Figure 10** an energy-harvesting

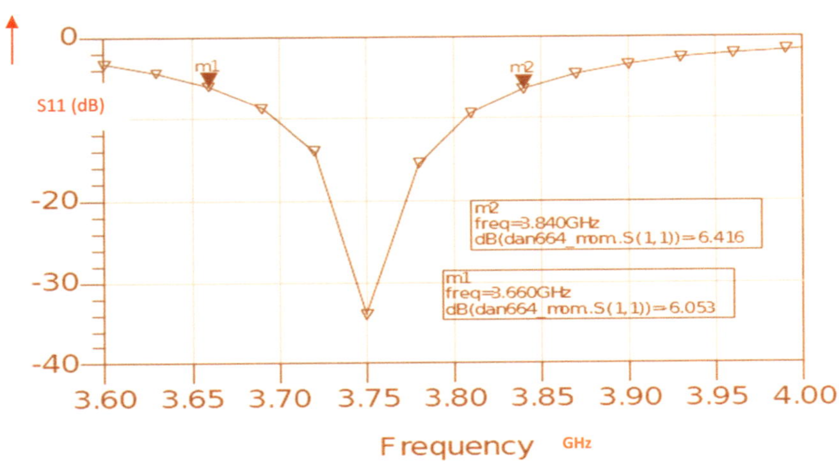

Figure 11.
Microstrip antenna with CSSR and energy-harvesting unit, S11 results.

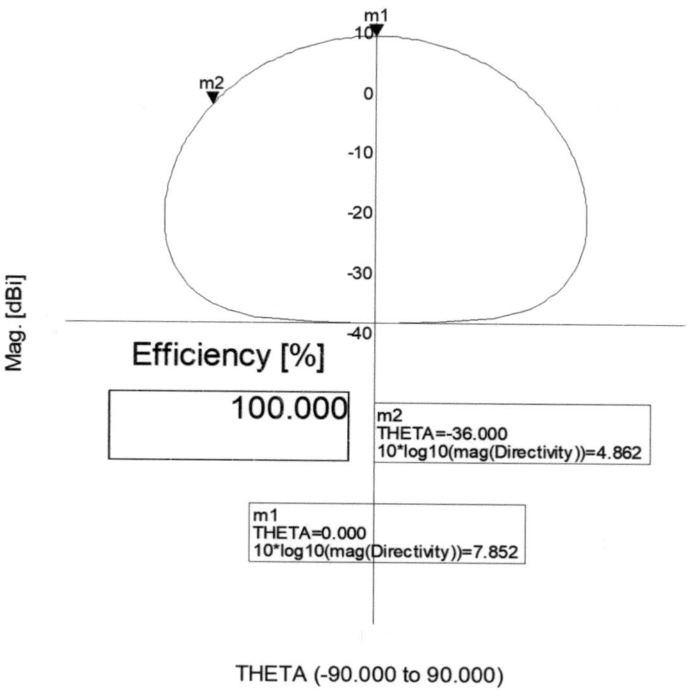

Figure 12.
Radiation pattern of the metamaterial antenna with CSRR.

unit is connected to the antenna and consists of a rectifying circuit, and a rechargeable battery, see Refs. [6–11, 25, 26]. The energy-harvesting unit and the antenna provide a self-powered green sensor. The rectifier diode converts RF energy, AC energy, to direct current (DC energy). The antenna radiation pattern is shown in **Figure 12**. This antenna may be used as green radiating elements in MIMO arrays, communication systems, and wearable medical devices. The antennas presented in this chapter were designed and optimized using ADS software [12].

Figure 11 presents S11 results of the antenna with CSSR on the human body. The patch resonant frequency is lower by 5% than the resonant frequency of the same antenna without CSSR. Printed antennas may be used as MIMO radiating element arrays. MIMO arrays are presented in Refs. [1–5, 27].

5. Conclusions

Printed antennas may be used as MIMO radiating element arrays. Microstrip antennas, double-layer patches, and metamaterial-printed antennas are compact efficient radiating elements that can be used as radiating elements in MIMO arrays, communication systems, and wearable medical devices.

The book will discuss and present new developments and innovation in several major topics in MIMO antennas and systems. The gain and directivity of the patch antenna without CSSR is lower by 2.5 dB than the patch antenna with CSSR. The metamaterial patch gain is around 7.7dBi. The antenna directivity is 8. The antenna efficiency is around 95%. There is a good agreement between measured and computed results.

Author details

Albert Sabban
ORT Braude College of Engineering, Karmiel, Israel

*Address all correspondence to: avi_111@netvision.net.il

References

[1] Larsson EG. Very Large MIMO Systems: Opportunities and Challenges. Mar 2012. Available from: http://www.kth.se/polopolyfs/1.303070!/Menu/general/columncontent/attachment/LargeMIMO.pdf

[2] Ngo HQ, Larsson EG, Marzetta TL. Energy and spectral efficiency of very large multiuser MIMO systems. IEEE Transactions on Communications. Apr 2013;**61**:1436-1449

[3] Li GY, Xu Z-K, Xiong C, Yang C-Y, Zhang S-Q, Chen Y, et al. Energy-efficient wireless communications: Tutorial, survey, and open issues. IEEE Wireless Communications Magazine. Dec 2011;**18**:28-35

[4] Xiong C, Li GY, Zhang S, Chen Y, Xu S. Energy- and spectral-efficiency tradeoff in downlink OFDMA networks. IEEE Transactions on Wireless Communications. Nov 2011;**10**:3874-3886

[5] Rusek F, Persson D, Lau BK, Larsson EG, Marzetta TL, Edfors O, et al. Scaling up MIMO: Opportunities and challenges with very large arrays. IEEE Signal Processing Magazine. Jan 2013;**30**:40-46

[6] Sabban A. Wideband RF Technologies and Antenna in Microwave Frequencies. USA: Wiley Sons; July 2016

[7] Sabban A. Novel Wearable Antennas for Communication and Medical Systems. New York, USA: Taylor and Francis Group; Oct 2017

[8] Sabban A. Wearable Systems and Antennas Technologies for 5G, IOT and Medical Systems. CRC Press, Taylor and Francis Group; Dec 2020. ISBN 9780367409135

[9] Sabban A. Wearable Communication Systems and Antennas, Second Edition: Design, Efficiency, and Miniaturization Techniques. Bristol, UK: IOP Publication; June 2022

[10] Sabban A, Gupta KC. Characterization of radiation loss from microstrip discontinuities using a multiport network modeling approach. IEEE Transactions on Microwave Theory and Techniques. Apr 1991;**39**(4):705-712

[11] U.S Patent. Inventors: Albert Sabban, Dual Polarized Dipole Wearable Antenna. USA; 19 Jun 2012. U.S Patent number: 8203497

[12] Software, Keysightt. Available from: http://www.keysight.com/en/pc-1297113/advanced-design-system-ads?cc=IL&lc=eng

[13] Sabban A. Microstrip antenna arrays. In: Nasimuddin N, editor. Microstrip Antennas. London, UK: InTech; 2011. pp. 361-384. Available from: http://www.intechopen.com/articles/show/title/microstrip-antenna-arrays. ISBN: 978-953-307-247-0

[14] Werber D, Schwentner A, Biebl EM. Investigation of RF transmission properties of human tissues. Advances in Radio Science. 2006;**4**:357-360

[15] Gupta B, Sankaralingam S, Dhar S. Development of wearable and implantable antennas in the last decade. In: Microwave Mediterranean Symposium (MMS). Guzelyurt, Turkey; Aug 2010. pp. 251-267

[16] Thalmann T, Popovic Z, Notaros BM, Mosig JR. Investigation and design of a multi-band wearable antenna. In: 3rd

European Conference on Antennas and Propagation. Berlin, Germany: EuCAP; 2009. pp. 462-465

[17] Salonen P, Rahmat-Samii Y, Kivikoski M. Wearable antennas in the vicinity of human body. In: IEEE Antennas and Propagation Society Symposium. Vol. 1. Monterey, USA; 2004. pp. 467-470

[18] Kellomaki T, Heikkinen J, Kivikoski M. Wearable antennas for FM reception. In: First European Conference on Antennas and Propagation. Hague, Netherlands: EuCAP; 2006. pp. 1-6

[19] Sabban A. Wideband printed antennas for medical applications. In: APMC 2009 Conference. Singapore; 12/2009

[20] Lee Y. Antenna Circuit Design for RFID Applications. Microchip Technology Inc., Microchip AN 710c

[21] Sabban A. Low Visibility Antennas for Communication Systems. N.Y, USA: Taylor and Francis Group; 2015

[22] Sabban A. A New Wideband Stacked Microstrip Antenna. I.E.E.E Antenna and Propagation Symp., Houston, Texas, U.S.A. June 1983

[23] Sabban A. PhD Thesis, Multiport Network Model for Evaluating Radiation Loss and Coupling among Discontinuities in Microstrip Circuits. Boulder, USA: University of Colorado; Jan 1991

[24] Rogers J, Plett C. Radio Frequency Integrated Circuit Design. Boston, MA, USA: Artech House; 2003

[25] Pendry JB, Holden AJ, Stewart WJ, Youngs I. Extremely low frequency plasmons in metallic mesostructures. Physical Review Letters. 1996;**76**:4773-4776

[26] Zhu J, Eleftheriades GV. A compact transmission-line metamaterial antenna with extended bandwidth. IEEE Antennas and Wireless Propagation Letters. 2009;**8**:295-298

[27] Zuo J et al. Multicell multiuser massive MIMO transmission with downlink training and pilot contamination precoding. IEEE Transactions on Vehicular Technology. 2016;**65**:6301-6314

Section 2

MIMO Antennas Systems Technologies

Chapter 2

Green Radio Frequency (RF) Technologies and Systems for MIMO Antennas Systems

Albert Sabban

Abstract

Multiple Input Multiple Output (MIMO) systems use multiple antennas. In this chapter, an RF module with eleven-notch antennas is presented. The RF module may operate as a phased array direction-finding system or as a MIMO antenna system. The antennas are wideband 6–18 GHz notch antennas. Moreover, a Microwave Integrated Circuits (MIC) X-band monopulse transceiver is presented in this chapter. The transceiver consists of a low-noise amplifier, mixer, attenuator, SP3T, limiter, and circulator. MIMO is employed in wideband RF systems where it is important not to have interference from RF systems. MIMO technology is used for Wi-Fi devices and cellular phones, 4G Long-Term Evolution (LTE) and 5G generation technology, broadcast TV production, law enforcement communication, and government applications. MIMO can be employed in WLANs, wireless local area networks. Green RF technologies, MIC, and Monolothic Microwave Integrated Circuits (MMIC) are presented in this chapter. Wideband passive and active RF modules and systems are presented in this chapter. Design rules and considerations of green electronic technologies in the development of green MIMO antenna systems are discussed in this chapter. RF and MIMO antenna systems should use green materials and green renewable energy.

Keywords: MIMO antennas, RF technologies, green technologies, MMICs, phased arrays, direction-finding systems, monopulse transceivers

1. Introduction

This chapter presents green RF technologies and systems for MIMO antenna systems. The development of efficient active devices can significantly reduce the total amount of energy consumption of communication and medical systems. Moreover, the usage of green renewable energy and green materials, when we develop RF components for the MIMO antenna system, will significantly reduce climate and environmental changes.

The popular microwave technologies for wearable communication systems are:

- MIC: Microwave Integrated Circuits
- MMIC: Monolithic Microwave Integrated Circuits

- MEMS: Micro-Electro-Mechanical Systems
- LTCC: Low Temperature Co-fired ceramic

It is better to design passive devices such as filters, couplers, and power dividers by using MIC technology. However, the production of miniature and low-cost active modules in mass production is achieved by using MMIC design. We can use the advantages and disadvantages of each technology in the development of RF modules. Microwave Integrated Circuits (MIC) technology allowed us to develop compact microwave modules. Active and passive modules can be assembled on the same substrate in Microwave Integrated Circuits. The invention of MMICs (Monolithic Microwave Integrated Circuits) technology led to the development of miniature RF systems. MMICs are usually fabricated on silicon and GaAs substrates. Microwave MMIC devices have small dimensions and low volume. For example, the dimensions of a 30 dBm, 35 GHz, power amplifier are around 4 × 3 × 0.1 mm. GaN transistors can operate at much higher temperatures and work at much higher voltages than GaAs transistors. Modern power amplifier MMICs are fabricated on gallium nitride. Indium phosphate, INP, provides better electrical performance than GaAs MMICs in terms of gain, low noise, and higher cutoff frequency. INP MMICs are more expensive due to the smaller wafer size. Small wearable communication and healthcare systems may be evaluated only by using modern RF technologies such as MMIC, MEMS, and LTCC technologies. RF module design is discussed in Refs. [1–5]. Wireless communication and the biomedical industry at RF and mm-wave frequencies are continuously growing. RF modules such as mixers, front-end couplers, filters, limiters, power amplifiers, printed MIMO antennas, and other passive components are important modules in personal wearable communication gadgets and MIMO antenna systems, see Refs. [1–11]. The electrical specifications of the RF modules determine if the system will achieve the required specifications. Minimization of the size and weight of RF devices is achieved by using MIC, MMIC, MEMS, and LTCC technologies. However, an accurate design and fabrication process is required to integrate MIC and MMIC components and modules. It is impossible to tune mm-wave RF devices in the fabrication process. We should use powerful electromagnetic design software to achieve accurate design of microwave MMIC and MIC devices and MIMO antennas. Accurate design of RF modules, mm-wave antennas, and transceivers is crucial in the development of low-cost communication smart devices.

MIMO systems are described in Refs. [12, 13]. In Ref. [13], the transmitted power rate proportional to the number of antennas at the base station is discussed. RF and MIMO antenna systems should use green materials and efficient green renewable energy. The energy efficiency of MIMO systems is discussed in Refs. [14–16]. To achieve a green environment, it is crucial to use efficient green renewable energy. Green electronic technologies and antennas are presented in Refs. [17–20].

Antennas with energy harvesting units that can be used in MIMO antenna systems are presented in Refs. [19, 21, 22]. A massive MIMO system is presented in Ref. [23].

Green renewable energy, green MIMO systems, and green materials are crucial in the development of RF components, as presented in this chapter. The usage of green technologies will considerably reduce air pollution, climate disasters, and environmental changes.

2. Examples of MIC and MMIC modules

An MIC X-band monopulse transceiver is shown in **Figure 1**. The transceiver consists of a low-noise amplifier, mixer, attenuator, SP3T, limiter, and circulator.

2.1 Green electronic technologies considerations in the design of MIMO systems

We should be aware of the design rules and considerations of green electronic technologies in the development of green MIMO antenna systems. We should use

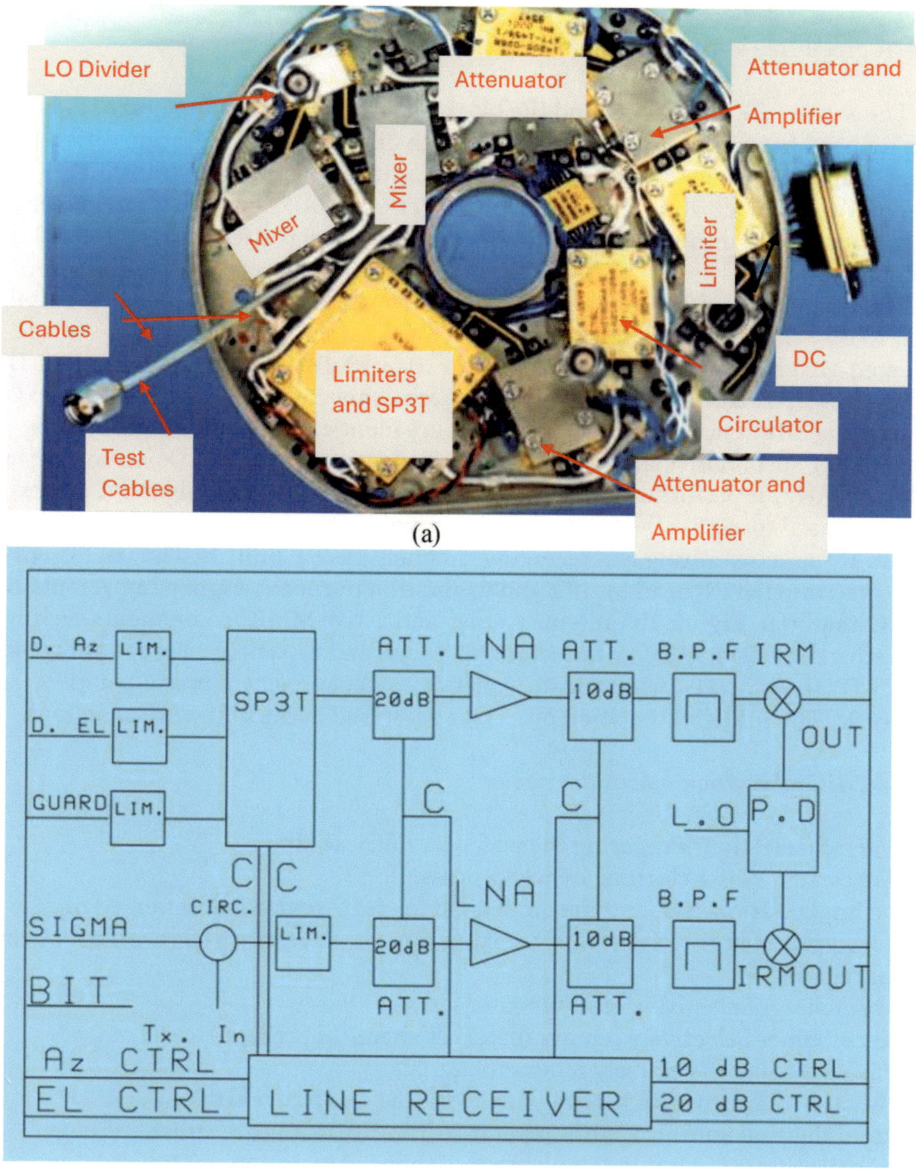

(b)

Figure 1.
a. Photo of a MIC X-band RF head. b. X-band transceiver block diagram.

green materials and energy-efficient amplifiers. The metal that should be used is aluminum, which is a green metal. A green-free lead solder should be used. The components that should be used to produce RF modules for MIMO antenna systems that are approved by RoHS standards.

2.2 Green technologies applied in MIMO antenna systems development

- **Green materials** should be **used**
- **Energy-efficient amplifiers** should be **used**
- A green free lead solder should be used

The X-band RF head photo and the block diagram are shown in **Figure 1a, b**. The RF head has five input ports: sum, azimuth difference and elevation difference, guard, and Built-In-Test (BIT). The transmitting channel can transmit a 60 W CW signal or 400 W pulse peak power. The sum RF signal, the transmitting channel, is routed through a circulator limiter and a digital attenuator with 0–20 dB attenuation level.

The output port of the digital attenuator is connected to a very low-noise amplifier with 26 dB gain and 1.2 dB noise figure. The SUM RF signal is routed to the Image Rejected Mixer via a 10 dB digital attenuator (0 or 10 dB) and a band-pass filter. The difference (ΔAz, ΔEl) antenna output ports are connected via a limiter to an SP3T switch. The difference (ΔAz, ΔEl) channels are identical to the sum channel. The guard input port is also connected to the SP3T switch via a limiter. Down conversion to IF (60 MHz) is performed by a 10.64 GHz local oscillator (LO) signal. A power divider is used to supply the local oscillator signal to the image reject mixer (IRM) mixers of the transmitting and receiving channels. An RF Built-In-Test (BIT) signal is injected into the RF head for BIT and calibration purposes. **Figure 2a** presents an MMIC amplifier. **Figure 2b** presents passive and active MMIC components such as field effect transistor (FETs), inductors, resistors, and capacitors. Pseudo-morphic HEMT (PHEMT) on GaAs substrate, 0.15 micron technology, is presented in **Figure 2c**. The MMIC fabrication process is described in the following paragraph.

2.2.1 MMIC fabrication process description

Wafer fabrication- Preparing the wafer for fabrication.

Wet cleans- Wafer cleaning by wet process.

Ion implantation- Dopants are embedded to create regions of increased or decreased conductivity. Selectively implant impurities. Create p- or n-type semiconductor regions.

Dry etching- Selectively remove materials.

Wet etching- Selectively remove materials chemical process.

Plasma etching- Selectively remove materials.

Thermal treatment- High-temperature process to remove stress.

Rapid thermal anneal- High-temperature process to remove stress.

Fu-mace anneal- After ion implantation, thermal annealing is required. Furnace annealing may take minutes and cause too much diffusion of dopants for some applications.

Oxidation- Substrate oxidation. Dry oxidation- $Si + O_2 \rightarrow SiO_2$.

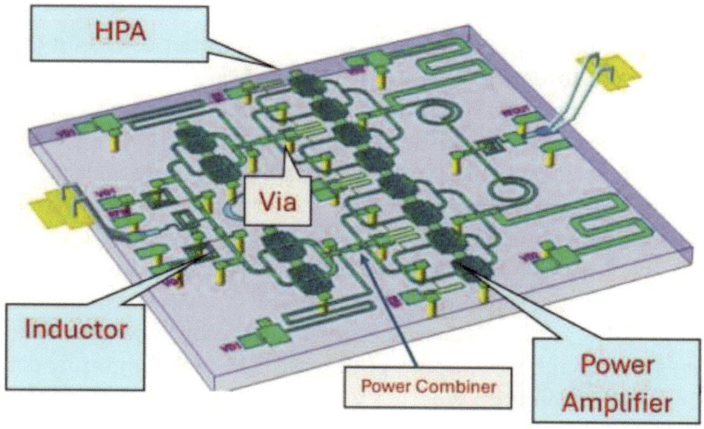

(a)

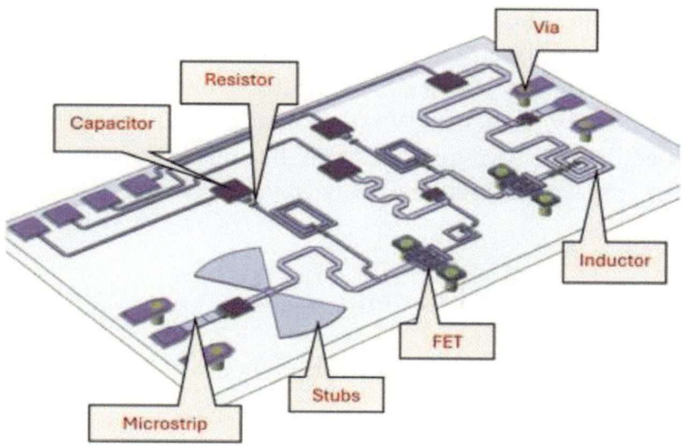

(b)

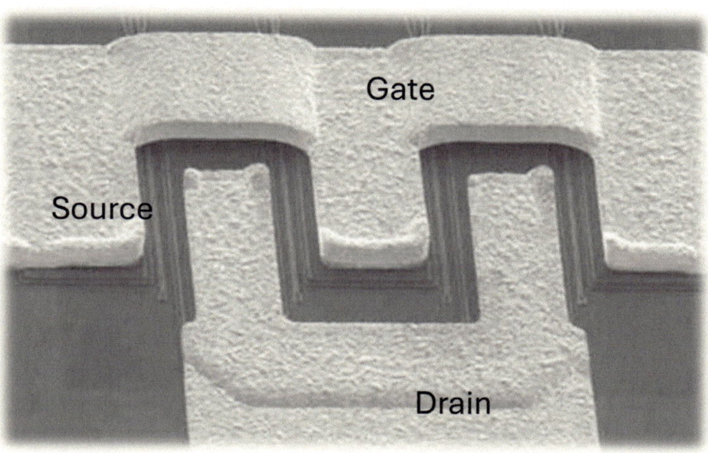

(c)

Figure 2.
MMIC technology a. Power amplifier b. Passive and active components c. 0.15 micron PHEMT on GaAs substrate.

Wet oxidation Si $+2H_2O \rightarrow SiO_2 + 2H_2$.

Chemical Vapor Deposition (CVD)- Chemical vapor deposited on the wafer. Pattern defined by photoresist.

Physical Vapor Deposition (PVD)- Vapor produced by evaporation or sputtering Deposited on the wafer. Pattern defined by photoresist.

Molecular Beam Epitaxy (MBE)- A beam of atoms or molecules produced in high vacuum. Selectively grow layers of materials. Pattern defined by photoresist.

Electroplating- Electromechanical process used to add metal.

Chemical Mechanical polish (CMP) wafer testing- Electrical test of the wafer.

Final process. Wafer back-grinding. Die preparation. Wafer mounting. Die cutting.

Lithography- Lithography is the process of transferring a pattern onto the wafer by selectively exposing and developing photoresist. Photolithography consists of four steps; the order depends on whether we are etching or lifting off the unwanted material.

Contact lithography- A glass plate is used that contains the pattern for the entire wafer. It is literally laid against the wafer during exposure of the photoresist. In this case, the entire wafer is patterned in one shot.

Electron-beam lithography is a form of direct-write lithography. Using E-beam lithography, you can write directly to the wafer without a mask. Because an electron beam is used, rather than light, much smaller features can be resolved.

Exposure can be done with light, UV light, or an electron beam, depending on the accuracy needed. E-beam provides much higher resolution than light because the particles are bigger (greater momentum), the wavelength is shorter.

3. Applications for MMIC components

Several microwave components are fabricated by using MMIC technology.

- Mixers - balanced, Star, sub-harmonic
- Amplifiers, LNAs, general, Power amplifiers, wideband power amplifiers, distributed TWA
- Switches - PIN, PHEMT, T/R matrix

3.1 MMIC components and modules

- Frequency multipliers - active, passive
- Modulators - QPSK, QAM (PIN, PHEMT)
- Multifunction - RX chip, TX chip, switched amp chip, LO chain
- Power amplifiers such as FET, BJT transistors, and High Electron mobility transistor (HEMT) transistors.

Low Noise Amplifiers, LNA, see Abbreviations at the end of the chapter.
Figure 3a,b present wideband, X-band, and Ka-band power amplifiers.

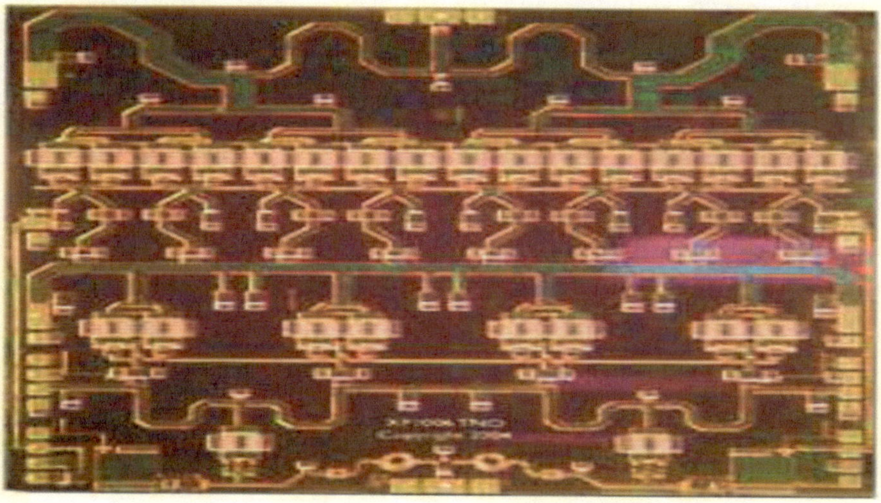

(a)

(b)

Figure 3.
MMIC power amplifiers. a. X-band amplifier b. Ka-band amplifier.

4. A wideband phased array for direction-finding systems and MIMO antenna system application

4.1 Introduction

The array network can operate as a direction-finding system or as a MIMO array. The wideband active phased array and a wideband receiving direction-finding system are presented in **Figure 4**. The system devices and block diagram are presented in **Figure 4**. The system consists of eleven receiving and transmitting antennas and modules for vertical scanning. The signals received in the six receiving antennas are amplified by LNA amplifiers and down-converted to a 70 MHz IF frequency. Input port number six provides the Local Oscillator (LO) signal to the module. The LO signal is amplified by the LNA amplifier and is routed to the Single Side Band (SSB) unit. A 70 MHz voltage-controlled oscillator supplies an RF signal to the SSB unit. The LO signal is amplified to 10 dBm. A wideband, six-way power divider supplies the LO signals to the mixers. The antennas are wideband, 6–18 GHz, notch

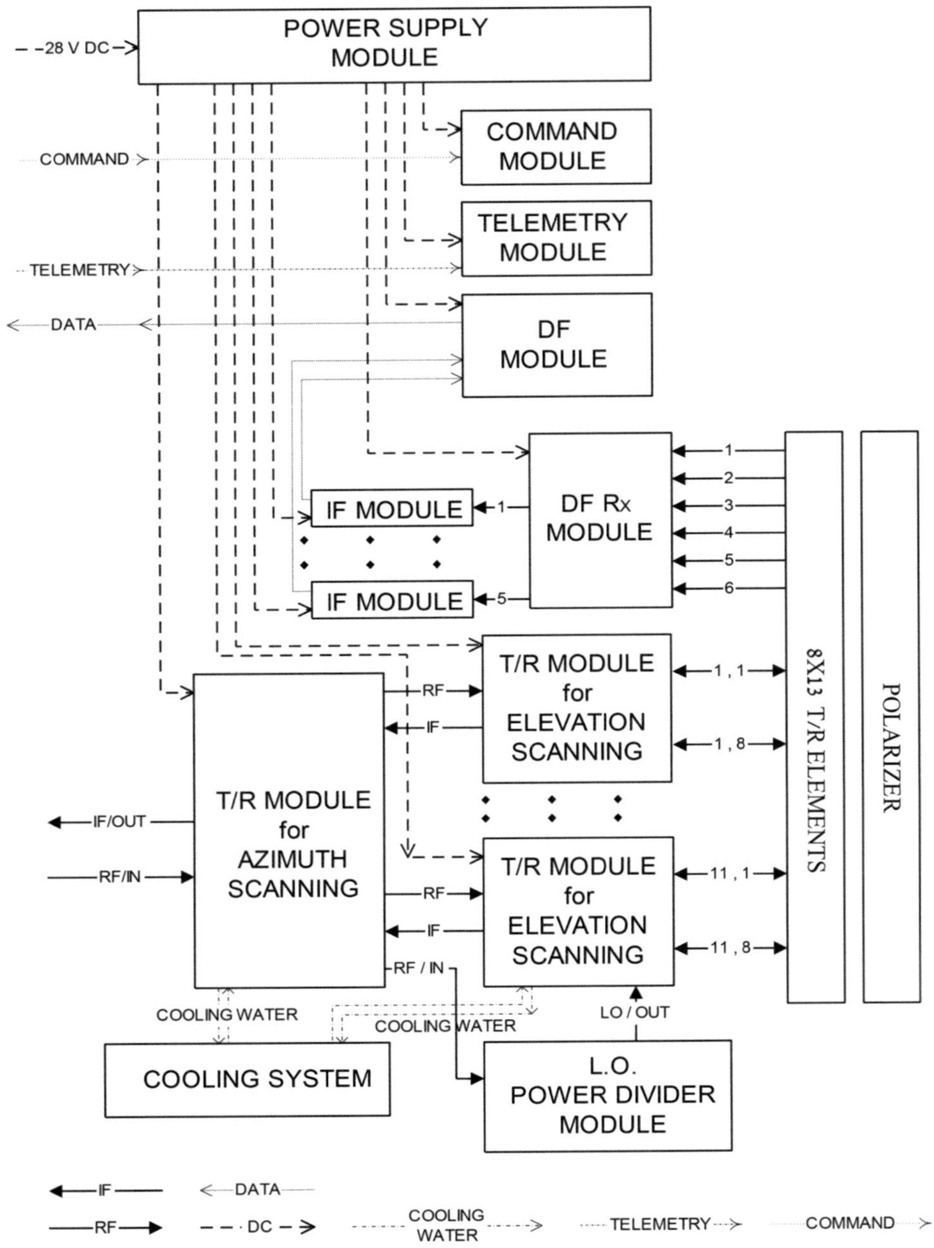

Figure 4.
Wideband 6–18 GHz System block diagram.

antennas, 6–18GHz, with voltage standing wave ratio (VSWR) better than 3:1. When this transceiver was developed, developers were not aware of the importance of green electronics to the environment. However, we did our best to use green materials and energy-efficient amplifiers.

4.2 Wideband receiving 6–18 GHz system

The receiving link consists of eleven receiving antennas and modules for vertical scanning. The receiving link components are LNA amplifiers and down converter mixers to 70 MHz IF frequency. The received signals in the six receiving ports are amplified by LNA amplifiers and down-converted to a 70 MHz IF frequency. Input port number six provides the Local Oscillator (LO) signal to the module. The LO signal is amplified by the LNA amplifier and is routed to the Single Side Band (SSB) unit. A 70 MHz voltage-controlled oscillator supplies an RF signal to the SSB unit. The LO signal is amplified to 10 dBm. A wideband six-way power divider supplies the LO signals to the mixers. The module dimensions are 279 × 130 × 15.5 mm. A cold water cooling system is used to cool the module hot spots.

4.2.1 The receiving link specifications

The 6–18 GHz receiving system specifications are listed in **Table 1**.

Parameter	Specifications
RF frequency	6–18 GHz
IF frequency	70 MHz
Gain	5 dB
Gain flatness	±2 dB
Phase flatness	±5 °
Noise figure	7 dB
Output power	0 dBm
Maximum input RF power	−7 dBm
Maximum input LO power	−7 dBm
60 MHz input RF power	10 dBm
IF ports	5
VSWR	3.1:1
Impedance	50 Ω
Dimensions	279 × 130 × 15.5 mm
Temperature	71°C÷-54°C
Weight	650 Gr.
DC power	19 W

Table 1.
6–18 GHz receiving system specifications.

4.3 6–18GHz receiving channel design

Figure 5 presents the block diagram of the 6–18 GHz direction-finding system.

The received signals are amplified by LNA amplifiers and down-converted to a 70 MHz IF frequency. The LNA amplifiers are packed MMICs with a 1 dB noise figure. The double-balance mixers are compact and have 7 dB insertion loss. The Local Oscillator (LO) signal is located in port number six of the module and is amplified by the LNA amplifier. The LO signal is amplified to 10 dBm. The LO signal is routed

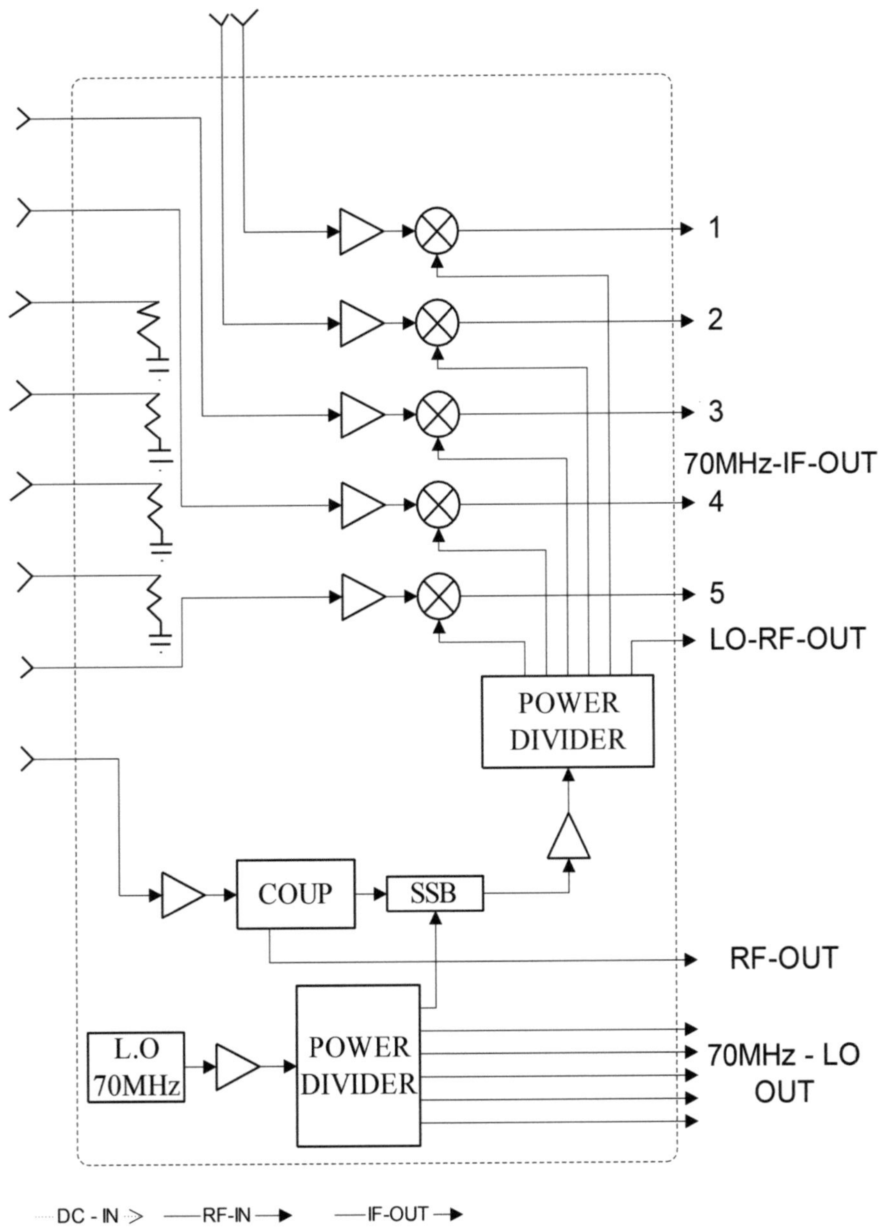

Figure 5.
Wideband 6–18 GHz receiving direction-finding system block diagram.

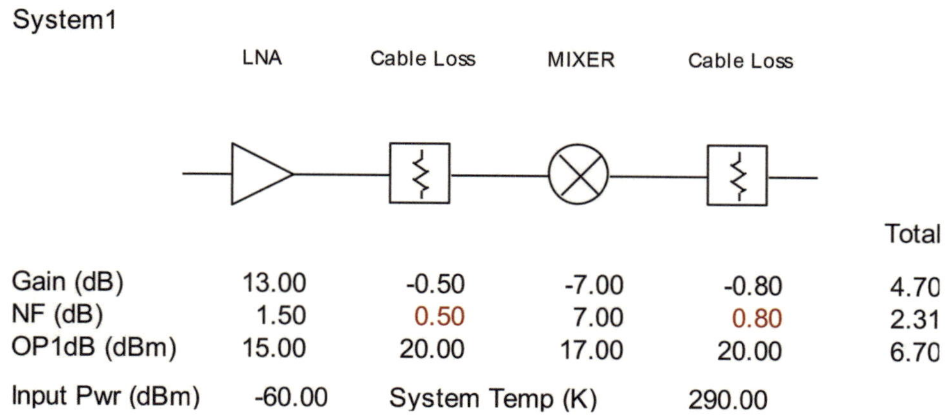

Figure 6.
Wideband receiving direction-finding system analysis design.

to the Single Side Band (SSB) unit. A 70 MHz VCO, voltage-controlled oscillator, generates the RF signal to the SSB unit. The LO signals to the mixers are routed via a six-way power divider. The IF and DC components are located on the back side of the module. The direction-finding channel analysis results are given in **Figure 6**. The analysis results prove that the system specifications can be achieved.

4.3.1 Cooling system

The direction-finding system cooling systems consist of heat sink pipelines. The cooling liquid flows via the heat sink pipelines and returns to the system through a cooling return connector.

4.3.2 DC unit

A twenty-four-pin DC connector supplies the DC voltages to the receiving system. The connector consists of seventeen DC pins and seven RF connectors. A list of the DC connector pins and DC power consumption is given in **Table 2**.

Pin	Function	Voltage	Current
1	VCO	15 V	0.9 A
2	LNA	6 V	0.1 A
3	LNA	6 V	0.1 A
4	LNA	6 V	0.1 A
5	IF AMP	15 V	0.1 A
6	Limiter	−15 V	
7		N.C	
8		N.C	
9		N.C	
10		N.C	
11	Limiter	15 V	

Pin	Function	Voltage	Current
12		N.C	
13		GND	
14		N.C	
15		N.C	
16		N.C	
17		N.C	

Table 2.
DC connector.

4.4 Measured results of the receiving channel

Measured results versus requirements of the receiving module are listed in **Table 3**.

A photo of the wideband receiving direction-finding system is shown in **Figure 7**.

A photo of the wideband receiving module with the module cover is shown in **Figure 8**. Absorbing material is glued on the cover to improve isolation between the

Parameter	Specifications	Measured results
RF frequency	6–18 GHz	6–18 GHz
IF frequency	70 MHz	70 MHz
Gain	4 dB	4 dB
Gain flatness	±2 dB	±2 dB
Phase flatness	±6 °	±6 °
Noise figure	7.5 dB	7.5 dB
Output power	0 dBm	0 dBm
Maximum input RF power	−7 dBm	−7 dBm
Maximum input LO power	−7 dBm	−7 dBm
60 MHz input RF power	10 dBm	10 dBm
1 dBc compression point	−10 dBm	−9 dBm
IP3	4 dBm	4 dBm
Spurious level	−40 dB	−45 dB
VSWR	3:1	3:1
Impedance	50 Ω	50 Ω
Dimensions	279 × 130 × 15.5 mm	279 × 130 × 15.5 mm
Temperature	71°C÷-54°C	71°C÷-54°C
Weight	700 Gr.	700 Gr.
DC power	20 W	20 W

Table 3.
Measured results versus requirements of the receiving module.

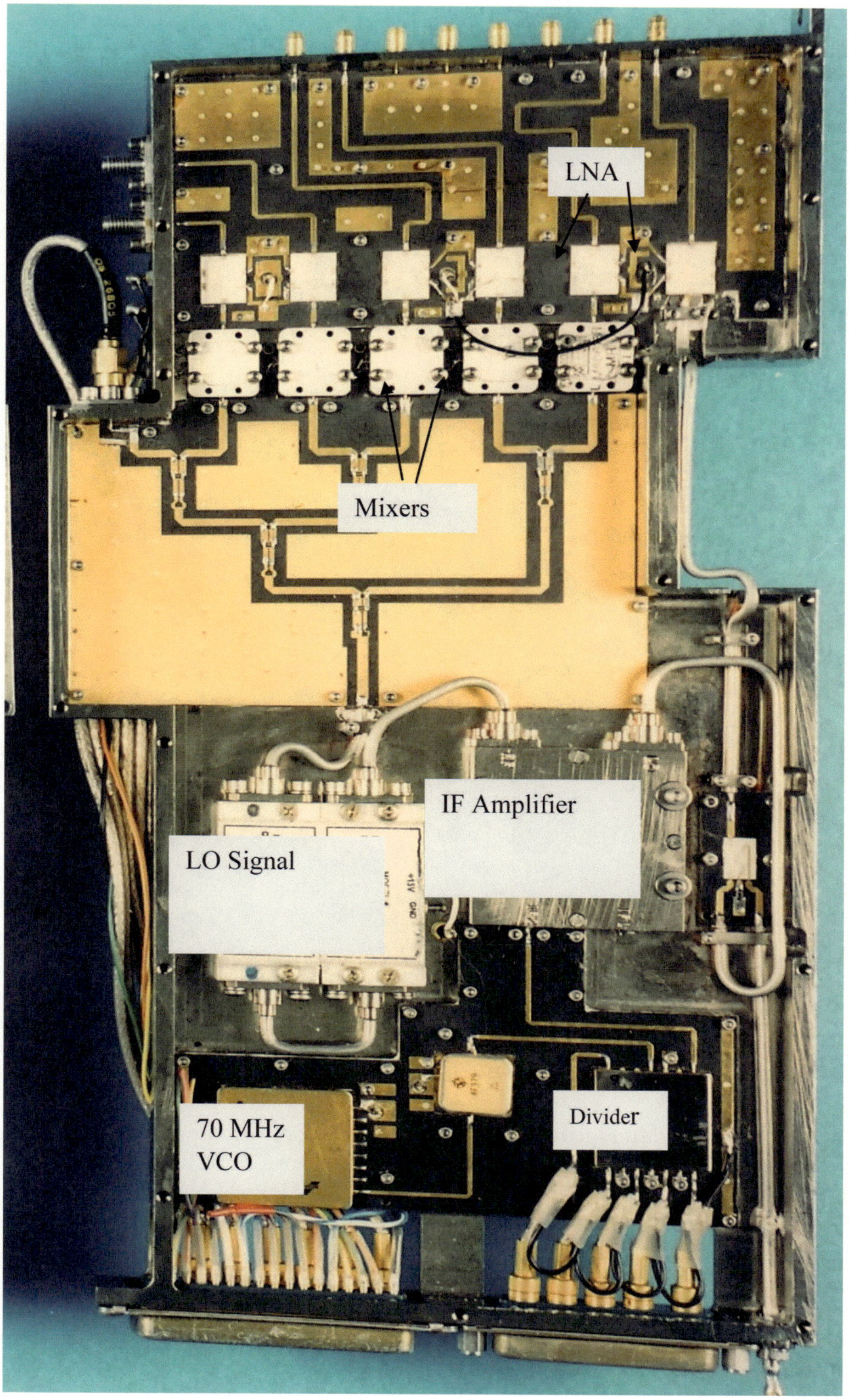

Figure 7.
Photo of the fabricated wideband 6–18 GHz receiving RF system.

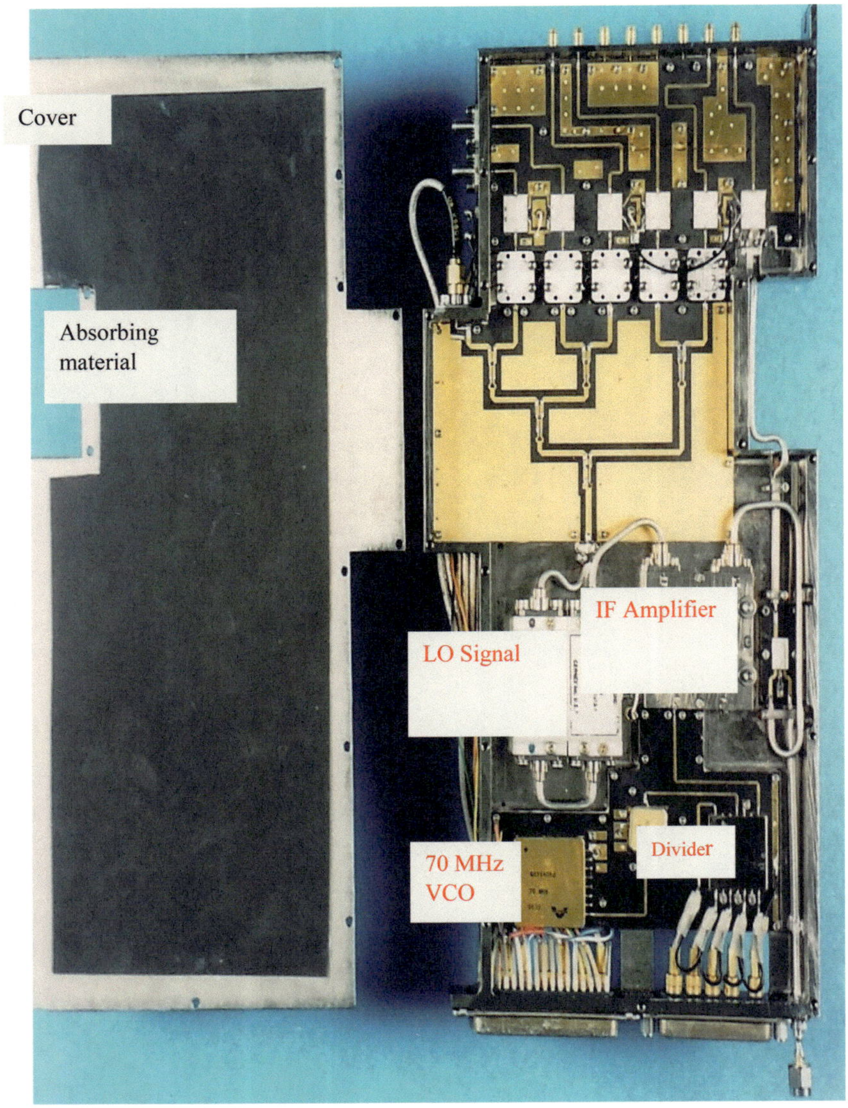

Figure 8.
Photo of the wideband 6–18 GHz receiving module with the module cover.

module components. In **Figure 9**, a photo of the Direction-Finding system interface and connectors is shown. The RF module dimensions are 279 × 130 × 15.5 mm. The compact dimensions were achieved by using compact-packed LNAs and mixers. However, a much more compact module may be designed by combining the LNAs, mixers, and power divider on the same MMIC module.

5. Green electronic technologies considerations in the design of the RF and MIMO antenna systems

RF and MIMO antenna systems should use green materials and renewable energy. The metal we used is aluminum, which is a green metal. The transceiver package

DOI: http://dx.doi.org/10.5772/intechopen.1009894

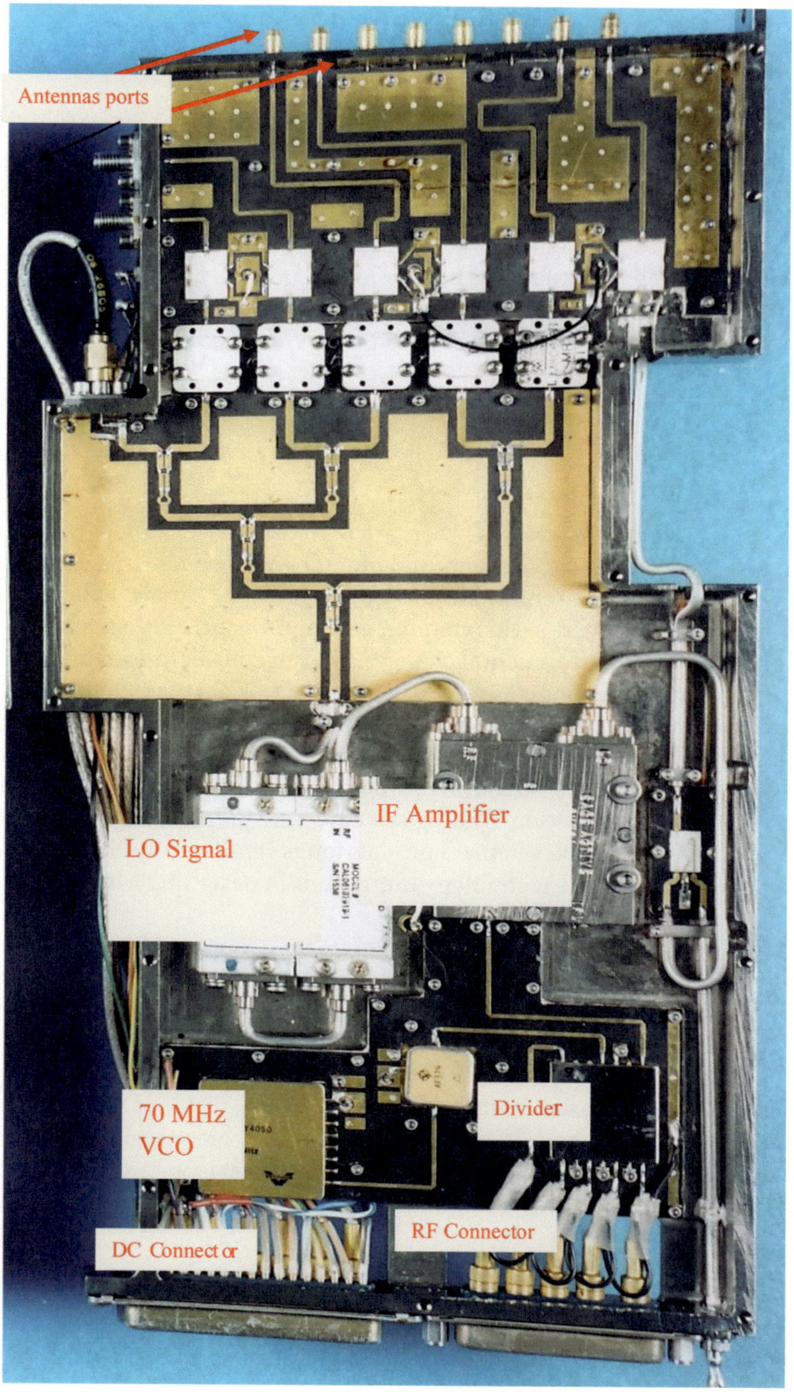

Figure 9.
Photo of the wideband 6–18 GHz receiving system interface.

and cover are produced from aluminum. A green-free lead solder was used. The components that were used to produce the transceiver were approved by the RoHS standards.

5.1 Green technologies applied in transceiver development

- Green materials were used
- Energy-efficient amplifiers were used
- A free lead green solder was used
- An efficient cooling system was used
- The cooling system is located on the amplifiers' backside
- RoHS standard was applied to all components

6. Conclusions

Green MIC and MMIC components and modules design were presented in this chapter. The usage of green renewable energy and green materials, when we develop RF components as presented in this chapter, will significantly reduce air pollution, climate disasters, and environmental changes. Design considerations and the evaluation of passive and active modules were described in this chapter.

In the design of RF modules, we may use the advantages of each technology. Miniature and low-cost modules in mass production are achieved by employing MMIC technologies. Green electronics technology considerations in the development of communication and direction-finding systems were presented in this chapter.

The cost of MMIC modules in mass production is significantly cheaper than MIC modules. However, there are technology limitations in designing MMIC LNAs, phase shifters, and high-power amplifiers.

RF passive and active components are used in energy harvesting modules. For example, power combiners, splitters, diodes, resistors, amplifiers, and other components. Antennas with energy harvesting units that can be used in MIMO antenna systems. A wideband, 6–18 GHz, receiving direction-finding system using MIC technology is presented in this chapter. Wideband six-way and five-way unequal power dividers were evaluated in this chapter.

In future communication systems, we can use energy harvesting units that provide green energy that may eliminate the need to replace batteries every day and the usage of power cords. An energy harvesting unit may be connected to the RF modules presented in this chapter.

Abbreviations

FET	field effect transistor
BJT	bipolar junction transistor
HEMT	high electron mobility transistor
PHEMT	pseudo-morphic HEMT
MHEMT	metamorphic HEMT
D-HBT	double hetero-structure bipolar transistor
CMOS	complementary metal-oxide semi-conductor

Author details

Albert Sabban
ORT Braude Engineering College, Israel

*Address all correspondence to: avi_111@netvision.net.il

References

[1] Rogers J, Plett C. Radio Frequency Integrated Circuit Design. Boston, MA: Artech House; 2003

[2] Maluf N, Williams K. An Introduction to Microelectromechanical System Engineering. Boston, MA: Artech House; 2004

[3] Sabban A. Microstrip antenna arrays. In: Nasimuddin N, editor. Microstrip Antennas. Rijeka, Croatia: InTech; 2011. pp. 361-384. ISBN: 978-953-307-247-0

[4] Sabban A. Applications of MM wave microstrip antenna arrays. In: ISSSE 2007 Conference; Montreal Canada. 2007

[5] Gauthier GP, Raskin GP, Rebiez GM, Kathei PB. A 94GHz micro-machined aperture-coupled microstrip antenna. IEEE Transactions on Antennas and Propagation. 1999;**47**(12):1761-1766

[6] Milkov MM. Millimeter-wave imaging system based on antenna-coupled bolometer [MSc thesis]. USA: UCLA; 2000

[7] de Lange G et al. A 3*3 mm-wave micro machined imaging array with sis mixers. Applied Physics Letters. 1999;**75**(6):868-870

[8] Rahman A et al. Micro-machined room temperature micro bolometers for MM-wave detection. Applied Physics Letters. 1996;**68**(14):2020-2022

[9] Mass SA. Nonlinear Microwave and RF Circuits. Boston, MA: Artech House; 1997

[10] Sabban A. Low Visibility Antennas for Communication Systems. New York, USA: Taylor & Francis Group; 2015

[11] Sabban A. Wearable Systems and Antennas Technologies for 5G, IOT and Medical Systems. New York, USA: CRC Press, Taylor & Francis Group; 2020. ISBN 9780367409135

[12] Larsson EG. Very Large MIMO Systems: Opportunities and Challenges. 2012. Available from: http://www.kth.se/polopolyfs/1.303070!/Menu/general/columncontent/attachment/LargeMIMO.pdf

[13] Ngo HQ, Larsson EG, Marzetta TL. Energy and spectral efficiency of very large multiuser MIMO systems. IEEE Transactions on Communications. 2013;**61**:1436-1449

[14] Li GY, Xu Z-K, Xiong C, Yang C-Y, Zhang S-Q, Chen Y, et al. Energy-efficient wireless communications: Tutorial, survey, and open issues. IEEE Wireless Communications Management. 2011;**18**:28-35

[15] Xiong C, Li GY, Zhang S, Chen Y, Xu S. Energy- and spectral-efficiency tradeoff in downlink OFDMA networks. IEEE Transactions on Wireless Communications. 2011;**10**:3874-3886

[16] Rusek F, Persson D, Lau BK, Larsson EG, Marzetta TL, Edfors O, et al. Scaling up MIMO: Opportunities and challenges with very large arrays. IEEE Signal Processing Magazine. 2013;**30**:40-46

[17] Sabban A. Advances in Green Technologies 2023. London: Intech Publication; 2023. DOI: 10.5722/Intechopen.100579

[18] Sabban A. Innovation in Green Computing Technologies and Computing Industry in 2021. London: Intech Publication; 2021

[19] Sabban A. Innovation in Global Green Technologies 2020. London: Intech Publication; 2020

[20] Sabban A. Green Ultra-Wideband RF Systems with Energy Harvesting Units for 5G and Medical Systems. London: Intech Publication; 2021

[21] Sabban A. Wearable self-powered sensors for health care, 5G, energy harvesting, and IOT systems. Biomedical Journal of Science and Technical Research. 2022;**41**:32374-32391. DOI: 10.26717/BJSTR.2022.41.006550

[22] Sabban A. Wearable circular polarized antennas for health care, 5G, energy harvesting, and IoT systems. Electronics. 2022;**11**(3):427. DOI: 10.3390/electronics11030427

[23] Zuo J et al. Multicell multiuser massive MIMO transmission with downlink training and pilot contamination precoding. IEEE Transactions on Vehicular Technology. 2016;**65**:6301-6314

Chapter 3

A Comprehensive Review of Massive MIMO Systems: Key Technologies, Challenges, and Future Directions

Bhawna Kalra, M.M. Sharma and Jaiverdhan

Abstract

The chapter provides a detailed examination of massive MIMO systems in wireless communications, focusing on the key benefits and challenges. It begins with an information-theoretic analysis, highlighting how the use of a large number of antennas in base stations can significantly improve spectral and energy efficiency. Practical challenges, such as channel estimation, detection, and precoding, are also addressed, with special emphasis on the issue of pilot contamination, caused by non-orthogonal pilot sequences from neighboring cells. Furthermore, this chapter evaluates the energy efficiency of massive MIMO systems and explores how the increased degrees of freedom can enhance single-carrier transmission efficiency. Finally, it explores both the challenges and opportunities that arise in the deployment of massive MIMO for future wireless communication systems. This chapter serves as a valuable resource for researchers and professionals interested in the advancements of wireless/mobile technologies toward 6G.

Keywords: massive MIMO systems, spectral efficiency, energy efficiency, time-division duplexing (TDD), channel estimation, pilot contamination, single-carrier transmission

1. Introduction

Over the past 20 years, multiple-input multiple-output (MIMO) technology has been extensively researched and incorporated into numerous wireless standards due to its ability to greatly enhance the capacity and reliability of wireless systems. Initially, research concentrated on point-to-point MIMO links, where two devices, each equipped with multiple antennas, communicate directly. However, recent research has shifted toward more practical multi-user MIMO (MU-MIMO) systems. In these systems, a base station (BS) with multiple antennas serves several single-antenna users simultaneously, allowing the multiplexing gain to be distributed among all users. This approach reduces the need for expensive equipment at the user end, as they can use relatively inexpensive single-antenna devices. Additionally, MU-MIMO systems benefit from multi-user diversity, making their performance less dependent on the propagation environment compared to point-to-point MIMO systems. Consequently, MU-MIMO has become a key component in communications standards such as 802.11 (WiFi), 802.16 (WiMAX), and LTE, and is increasingly being deployed worldwide. In many MIMO implementations, base stations typically use fewer than ten antennas, which, while improving spectral efficiency, result in relatively modest gains.

Recent advancements in wireless communication have led to the development of massive MIMO systems, also known as large-scale antenna systems (LSAS), which aim to achieve significant performance improvements and simplify signal processing requirements. In these systems, each base station (BS) is equipped with a significantly higher number of antennas—often 100 or more—compared to traditional MIMO systems [1, 2]. **Figure 1** illustrates a massive MU-MIMO network, where a large number of antennas at the BS serve multiple users simultaneously.

The theoretical foundation for the benefits of massive MIMO comes from random matrix theory, which suggests that as the number of antennas increases, the impact of uncorrelated noise and small-scale fading diminishes [2]. This means that in an ideal massive MIMO system, the quality of communication improves as the number of antennas grows, the number of users per cell becomes independent of the cell size, and the energy required to transmit each bit decreases toward zero. Moreover, massive MIMO systems allow for the use of simple linear signal processing techniques, such as matched-filter (MF) precoding and detection, which are sufficient to harness these performance gains effectively.

According to Ref. [2], under realistic propagation conditions, massive MIMO systems utilizing matched filter (MF)-based noncooperative techniques can theoretically achieve a data rate of 17 Mbps for each of 40 users within a 20 MHz channel. This applies to both uplink (reverse link) and downlink (forward link) communications, resulting in an average cell throughput of 730 Mbps and an overall spectral efficiency of 26.5 bits per second per Hertz (bps/Hz). Because the number of antennas at the base station (BS) is typically much greater than the number of users, these systems have a large number of degrees of freedom. This abundance allows for the efficient shaping of transmitted signals and interference nulling [3]. To implement such systems effectively, it is crucial to develop algorithms that manage complexity while harnessing these advantages.

Massive MIMO systems offer notable advantages in energy efficiency compared to traditional single-antenna systems. According to Ref. [4], in a massive MIMO setup, a

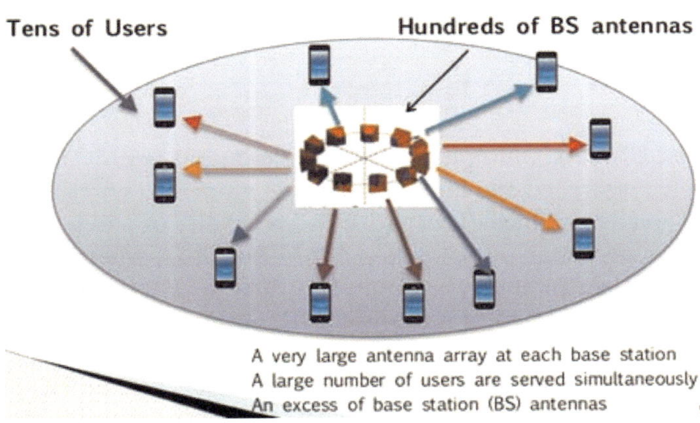

Figure 1.
Illustration of massive MU-MIMO systems.

single-antenna user can reduce its transmit power proportionally to the number of antennas at the base station (BS) when perfect channel state information (CSI) is available. If CSI is imperfect, the transmit power can be scaled down by the square root of the number of BS antennas while maintaining equivalent performance to a single-input single-output (SISO) system. This scaling leads to improved energy efficiency, which is increasingly crucial for future wireless networks facing growing energy consumption concerns [5, 6]. Additionally, with sufficient transmit power, massive MIMO can significantly extend operational range compared to single-antenna systems. Despite the fact that the analysis in Ref. [4] does not account for the power consumption of the radio front-end, massive MIMO remains a highly promising technology for enhancing the energy efficiency of future networks.

The recent observations about massive MIMO systems have led to a surge of research focused on understanding their signal processing and information theoretic implications. A review in Ref. [7] covers various aspects of massive MIMO, including fundamental information-theoretic gains, antenna and propagation considerations, and transceiver design. A subsequent tutorial [8] provides a brief update on recent developments. This chapter aims to offer a more thorough and detailed overview of the latest research on massive MIMO systems.

In Section 2, we explore the potential benefits of massive MIMO from an information-theoretic standpoint. Section 3 addresses challenges related to channel estimation and signal detection, while Section 4 delves into various transmit precoding schemes. Beyond the matched-filter (MF) precoder/detector, we also examine linear approaches such as minimum mean square error (MMSE) and zero-forcing (ZF) precoders/detectors, considering both single-cell and multi-cell coordinated processing scenarios. Section 5 discusses the pilot contamination problem, which arises from using non-orthogonal pilot sequences among users in different cells. In Section 6, we analyze the energy efficiency of massive MIMO systems and the concluding remarks provided in Section 7.

2. MIMO to massive MIMO

This section delves into the advantages of massive MIMO systems from an information-theoretic perspective. We begin by analyzing point-to-point MIMO systems, where the focus is on understanding how equipping each terminal with a large number of antennas can unlock significant performance improvements. This analysis sheds light on the potential benefits of increased antenna count at each communication endpoint.

Subsequently, we explore the performance of multi-user MIMO (MU-MIMO) systems. In these systems, a base station (BS) equipped with a large array of antennas serves multiple single-antenna users simultaneously. This scenario allows us to examine how massive MIMO can enhance communication efficiency and capacity in practical, multi-user settings.

The discussion in this section builds on findings from Refs. [7, 9, 10], which provide comprehensive insights into the theoretical and practical advantages of massive MIMO technology.

2.1 Point-to-point MIMO

We begin by examining point-to-point MIMO transmission, where the transmitter is equipped with N_t antennas, and the receiver has N_r antennas. Our analysis focuses on a narrowband, time-invariant channel with a deterministic and constant channel matrix H ϵ $C^{Nr \times Nt}$. To handle frequency-selective wideband channels, which can be transformed into multiple parallel flat-fading narrowband channels, orthogonal frequency-division multiplexing (OFDM) schemes are commonly employed [11]. In this context, the received signal vector y ϵ $C^{Nr \times 1}$ can be described as follows:

$$y = \sqrt{\rho}Hx + n \qquad (1)$$

In this setup, $x \in C^{Nr \times 1}$ denotes the transmit signal vector, while ?∈ $C^{Nr \times 1}$ represents the noise and interference. We assume that the total power of the transmit signal is normalized, meaning that $E[\|x\|^2]$ =1 and that the noise follows a zero-mean, circularly symmetric complex Gaussian distribution with an identity covariance matrix [I]. Under these conditions, the scalar value represents the transmit power. Assuming that the transmit signals are independent and identically distributed (I.I.D.) Gaussian and that perfect channel state information (CSI) is available at the receiver, the achievable rate at any given moment can be calculated as follows:

$$C = \log_2 \det\left(1 + \frac{\rho}{N_t} HH^T\right) bit/\sec/Hz \qquad (2)$$

When the propagation coefficients in the channel matrix H are normalized such that the trace of HH^T is equal to N_t N_r, where N_t and N_r are the number of antennas at the transmitter and receiver, respectively, the bounds on the channel capacity can be determined. This normalization ensures that the total power of the channel matrix is properly scaled. In Ref. [7], upper and lower bounds on the channel capacity are derived using Jensen's inequality, a mathematical tool that provides bounds for functions of random variables.

$$log_2 \det(1 + \rho N_r) \leq C \leq \min(N_t N_r) \log_2\left(1 + \frac{\rho \max(N_t N_r)}{N_t}\right) \qquad (3)$$

The actual achievable rate of a MIMO channel depends on how the singular values of HH^H are distributed. Among all possible channels with the same normalization, the highest achievable rate occurs when all singular values are equal, which corresponds to the upper bound given in Eq. (3). Conversely, the lowest rate is achieved when there is only one nonzero singular value, corresponding to the lower bound in Eq. (3). The optimal scenario, where the rate approaches the upper bound, occurs when all propagation coefficients in the channel matrix are I.I.D. In contrast, the worst-case scenario might involve line-of-sight (LOS) propagation, where the channel has limited diversity.

We will now examine two extreme cases in which either the number of transmit antennas N_t or the number of receive antennas N_t approaches infinity:

1. When $N_t > > N_t$ and N_t tends to ∞: In this scenario, as the number of transmit antennas grows infinitely large, the system's performance is analyzed under these conditions. is constant, i.e., N_t N_r N_t, the row vectors of H are asymptotically orthogonal, and hence we have

$$\left(\frac{HH^H}{N_t}\right) \approx I_{N_r} \qquad (4)$$

The achievable rate from (2) can be written as

$$c = N_t \log_2(1+\rho) bit/s/Hz \tag{5}$$

2. When $N_t >> N_t$ and N_r tend to ∞, apply the same derivation as in condition (1), we get

$$C = N_t \log_2\left(1+\frac{\rho N_r}{N_t}\right) bit/s/Hz \tag{6}$$

The results presented in Eqs. (5) and (6) highlight the benefits of using a large number of antennas in a MIMO system. However, these conclusions rely on the assumption that the rows or columns of the channel matrix \(H\) are asymptotically orthogonal, which is an idealized scenario regarding the propagation coefficients. In real-world situations, this level of orthogonality is optimistic. In environments with line-of-sight (LOS) propagation, as discussed in Ref. [7], the advantage of increased multiplexing gain diminishes, indicating that the performance improvements of adding more antennas are less pronounced in such cases.

2.2 Multi-use MIMO

Multi-user MIMO (MU-MIMO) systems offer the potential to achieve the promising multiplexing gains found in massive point-to-point MIMO systems while mitigating issues related to challenging propagation environments.

In an MU-MIMO setup, imagine a network consisting of L cells. Within each cell, there is a base station (BS) equipped with N antennas, and K single-antenna users are served. The channel coefficient from the k-th user in the I-th cell to the n-th antenna of the i-th base station is denoted as $h_{i,k,l,n}$. This coefficient represents a complex small-scale fading factor, which captures rapid fluctuations in signal strength due to the movement of objects or the environment. It is multiplied by an amplitude factor that accounts for geometric attenuation and large-scale fading. Geometric attenuation relates to the signal's reduction in strength due to the distance between the transmitter and receiver, while large-scale fading considers variations in signal strength due to factors such as shadowing from buildings or terrain.

$$h_{i,k,l,n} = g_{i,k,l,n}\sqrt{d_{i,k,l}} \tag{7}$$

Where

$g_{i,k,l,n}$ = complex small-scale fading.

$d_{i,k,l}$ = complex large-scale fading.

So, MU-MIMO systems leverage the benefits of massive MIMO technology to enhance performance across multiple users simultaneously, overcoming the limitations posed by unfavorable propagation conditions that can affect point-to-point MIMO setups.

3. Channel estimation and signal detection

This section delves into the processes of channel estimation and signal detection at the base station (BS) within a massive MIMO system. We start by exploring the various methods used for channel estimation in massive MIMO setups and discuss why the time division duplex (TDD) mode is typically preferred for these systems.

Channel estimation is a critical component in massive MIMO systems because it involves determining the channel state information (CSI), which is essential for accurate communication. In massive MIMO, where there are many antennas, the estimation process becomes more complex. We will cover different techniques used to estimate the channel, including those that leverage a large number of antennas to improve accuracy and efficiency.

A key reason why TDD mode is often assumed in massive MIMO systems is its suitability for channel estimation. TDD allows for the use of uplink (UL) pilot signals to estimate the channel, which can then be used for downlink (DL) communication. This mode is advantageous because it leverages the reciprocity of the channel—the channel characteristics are assumed to be the same for both uplink and downlink if the transmission frequencies are identical. This reciprocity simplifies the channel estimation process, making it more efficient compared to frequency division duplex (FDD) mode, where separate estimation procedures are needed for uplink and downlink.

Following the discussion on channel estimation, we turn our attention to signal detection methods at the BS. Signal detection involves interpreting the received signals to recover the transmitted data accurately. We will present both linear and nonlinear signal detection techniques. Linear methods include approaches such as matched-filter (MF) detection and zero-forcing (ZF) detection, which are often favored for their simplicity and ease of implementation. Nonlinear methods, such as minimum mean square error (MMSE) detection, are also discussed as they may offer improved performance in terms of error rates and overall signal quality.

3.1 Channel estimation

In traditional multiple-input multiple-output (MIMO) systems, the process of multi-user precoding in the downlink and detection in the uplink both rely heavily on having accurate channel state information (CSI) at the base station (BS). The resources needed for estimating the channel in a MIMO system—whether in terms of time or frequency—depend on the number of transmit antennas but do not depend on the number of receive antennas.

When frequency division duplexing (FDD) is employed, where the uplink and downlink operate on different frequency bands, the CSI for these two channels (uplink and downlink) is different due to the variation in frequency. To estimate the CSI for the uplink, the base station instructs all users to send distinct pilot sequences. The time required for transmitting these uplink pilots is not influenced by the number of antennas at the base station.

However, in FDD systems, obtaining CSI for the downlink is more complex and requires a two-step process. First, the base station transmits pilot symbols to all users. Then, each user estimates the downlink CSI and feeds it back to the base station. The time required to transmit these downlink pilot symbols is directly proportional to the number of antennas at the base station.

As the number of antennas at the base station increases significantly, this traditional method of downlink channel estimation in FDD systems becomes impractical. For instance, consider a scenario where there is a 1 ms coherence interval over a 100 kHz channel, which can accommodate the transmission of 100 complex symbols. If the base station has 100 antennas, and orthogonal pilot waveforms are used to estimate the channels for each antenna, the entire coherence interval would be

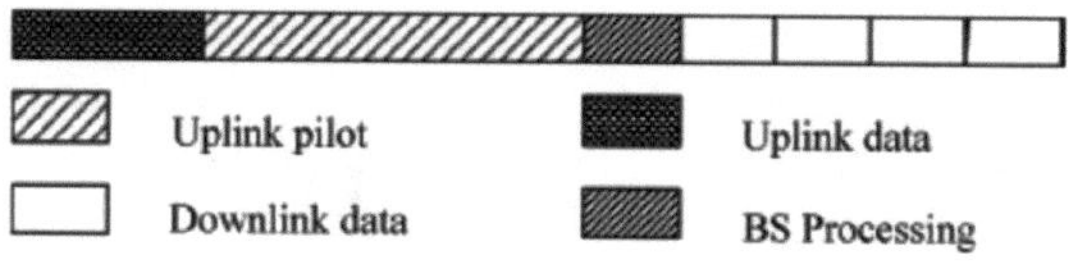

Figure 2.
Multi-user MIMO TDD protocol.

consumed by downlink training alone, leaving no time for actual data transmission. This illustrates the challenge of scaling the traditional downlink channel estimation approach as the number of antennas at the base station increases.

The channel estimation strategy in time division duplex (TDD) systems offers a viable solution to this challenge. By leveraging the principle of channel reciprocity, it is only necessary to estimate the channel state information (CSI) for the uplink. In Ref. [12], a specific TDD protocol, depicted in **Figure 2**, was introduced. This protocol works as follows: All users across all cells simultaneously transmit their uplink data signals. Subsequently, the users send out pilot sequences. These sequences are then used by the base stations (BSs) to estimate the CSI for the users within their respective cells. The estimated CSI is crucial as it allows the BSs to detect the uplink data and create beamforming vectors for downlink data transmission.

However, there is a complication due to the limited channel coherence time. The pilot sequences that users in neighboring cells utilize might no longer remain orthogonal to those within the cell. This situation gives rise to a pilot contamination issue, as described in Ref. [2]. This problem occurs when the pilot sequences overlap, causing interference and degrading the overall performance of the system.

Linear minimum mean square error (MMSE)-based channel estimation is a widely adopted method due to its ability to deliver near-optimal performance while maintaining low computational complexity. This technique is favored because it effectively minimizes the estimation error by considering both the signal and noise statistics, resulting in accurate channel estimation with minimal resource expenditure. Beyond the traditional MMSE estimation, an alternative approach based on compressive sensing is discussed in Ref. [13]. This method leverages the observation that the degrees of freedom in the physical channel matrix are significantly fewer than the number of potential-free parameters. In essence, compressive sensing capitalizes on the sparsity of the channel matrix, allowing for efficient estimation with fewer measurements. This approach is particularly beneficial in scenarios where the channel exhibits a sparse structure, enabling more efficient data acquisition and processing.

To further enhance the system's spectral efficiency, a novel time-frequency training sequence design is proposed in Ref. [14]. This design aims to optimize the training process by combining the advantages of both time-domain and frequency-domain estimation techniques. Traditional time-domain estimation is known for its simplicity but can suffer from frequency selectivity issues. On the other hand, frequency-domain estimation offers better performance in frequency-selective channels but can be computationally intensive. The proposed time-frequency training structure effectively integrates the strengths of both domains while mitigating their respective limitations, leading to improved channel estimation accuracy and overall system performance.

3.2 Signal detection

In massive MIMO systems, linear signal detectors such as the matched filter (MF), zero-forcing (ZF), and minimum mean square error (MMSE) detectors are often considered practical choices due to their low computational complexity. These detectors become increasingly effective as the number of antennas at the base station (BS) grows significantly larger than the number of users. In such scenarios, where the channel vectors from different users are independent, these linear detectors can asymptotically approach the system's capacity, making them highly efficient for massive MIMO environments.

Extensive research has been conducted to evaluate the performance of massive MIMO systems utilizing various linear receivers. For instance, a comparative analysis between the MMSE and MF receivers under realistic system conditions is presented in Ref. [15]. The study reveals that the MMSE receiver can achieve comparable performance to the MF receiver but with fewer antennas, particularly in environments where inter-cell interference is present. This advantage makes the MMSE receiver a more efficient choice in such scenarios, as it requires fewer resources to maintain high performance.

Further investigations have been carried out to understand the behavior of MMSE and ZF receivers when the ratio of the number of antennas to the number of users is bounded. In Refs. [16, 17], this scenario is analyzed for MMSE and ZF receivers, respectively. In Ref. [16], an expression for the asymptotic signal-to-interference-plus-noise ratio (SINR) of the MMSE receiver in a single-cell system is derived, taking into account a bounded ratio of antennas to users. The study also distinguishes between two types of MMSE receivers: the optimal MMSE receiver, which considers varying transmit power levels from different users, and a suboptimal MMSE receiver, which assumes equal transmit power among users.

Similarly, Ref. [17] delves into the performance metrics of ZF receivers, such as the exact data rate, symbol error rate, and outage performance. These metrics provide a comprehensive understanding of how ZF receivers operate under different conditions. In addition to centralized MIMO systems, Ref. [18] explores the sum rate performance of ZF receivers in distributed MIMO systems. The study establishes lower and upper bounds on the sum rate, providing valuable insights into the performance limits of ZF receivers in distributed setups.

Regarding the computational complexity of these receivers, a rough estimation shows that both ZF and MMSE receivers have a complexity order of $O(NK + NK^2 + K^3)$, where N represents the number of antennas and \(K\) the number of users. This complexity analysis highlights the balance between performance and computational effort required for implementing these linear detectors in massive MIMO systems.

In the realm of massive MIMO systems, while linear detection methods such as matched filter, zero-forcing, and MMSE detectors are popular for their low computational complexity, nonlinear detection techniques offer the potential for even better performance. However, this enhanced performance comes at the expense of significantly higher computational complexity. The primary challenge with nonlinear detectors is managing this complexity, making it feasible for practical deployment in massive MIMO environments.

One of the key focuses of recent research has been on reducing the computational burden associated with nonlinear detection methods, allowing them to be more effectively integrated into massive MIMO systems without overwhelming processing resources. Addressing this challenge, some innovative approaches have been proposed to strike a balance between the benefits of nonlinear detection and the need for manageable complexity.

A notable example is found in Ref. [19], where a block-iterative generalized decision feedback equalizer (BI-GDFE) is introduced as a nonlinear detection technique. The BI-GDFE is designed to improve the performance of massive MIMO systems by iteratively refining the detection process, thereby enhancing the quality of signal reception. This method is particularly advantageous in scenarios with high interference or when dealing with complex channel conditions, where linear detectors might fall short in performance. The BI-GDFE operates by leveraging a decision feedback mechanism, which iteratively processes blocks of data to refine the signal estimation. By feeding back previous decisions into the detection process, the BI-GDFE can correct errors that might occur in the initial detection stages, leading to a more accurate estimation of the transmitted signals [20]. This iterative process, while computationally intensive, enables the BI-GDFE to achieve superior performance compared to linear detectors, especially in challenging environments.

To assess the effectiveness of the BI-GDFE, its asymptotic signal-to-interference-plus-noise ratio (SINR) performance is evaluated. The asymptotic SINR is a critical metric that indicates how well the detector can separate the desired signal from interference and noise as the system size scales up. The evaluation of the BI-GDFE's SINR performance demonstrates its potential to significantly enhance signal detection in massive MIMO systems, making it a promising candidate for scenarios where maximum performance is required. Despite its benefits, the higher computational complexity of the BI-GDFE and similar nonlinear detectors remain a concern. Therefore, ongoing research continues to explore methods to optimize these techniques, aiming to reduce their complexity without sacrificing the performance gains they offer. By doing so, nonlinear detection methods could become more practical and widely adopted in massive MIMO systems, unlocking new levels of performance in wireless communications.

4. Precoding

This section focuses on the concept of precoding at the base station (BS). We begin by exploring basic precoding techniques and then expand the discussion to include multi-cell precoding strategies. Additionally, we will address some practical challenges associated with implementing precoding in real-world systems.

In traditional MIMO systems, both linear and nonlinear precoding techniques are employed to optimize signal transmission. Nonlinear methods, such as dirty-paper coding (DPC), vector perturbation (VP), and lattice-aided approaches, are known for their superior performance compared to linear techniques. These nonlinear methods effectively mitigate interference and enhance signal quality, making them ideal for complex scenarios. However, this improved performance comes at the cost of significantly higher computational complexity, making them more challenging to implement, especially as the system size grows.

As the number of antennas at the BS increases, particularly in massive MIMO systems, the dynamics change. Research has shown that linear precoders, such as matched filter (MF) and zero-forcing (ZF), become nearly optimal in such scenarios. This near-optimality occurs because the large number of antennas allows the BS to effectively manage interference and improve signal quality, reducing the gap between the performance of linear and nonlinear precoding techniques.

Given this shift in performance dynamics, it becomes more practical to utilize low-complexity linear precoding techniques in massive MIMO systems. These linear methods, despite their simplicity, can achieve performance levels close to those of more complex nonlinear approaches when the BS is equipped with a large number of antennas. This makes linear precoding not only more feasible in terms of implementation but also more attractive from a practical standpoint, particularly in large-scale deployments where computational resources may be limited.

After discussing basic precoding methods, the conversation extends to multi-cell precoding. In multi-cell scenarios, the challenge is not only to optimize signal transmission within a single cell but also to coordinate across multiple cells to minimize interference and maximize overall system performance. Multi-cell precoding strategies aim to address these challenges, ensuring that the benefits of massive MIMO systems can be realized even in more complex, real-world environments.

Finally, we consider some practical issues related to precoding, such as hardware limitations, channel estimation errors, and the impact of imperfect synchronization across cells. These factors can significantly influence the performance of precoding techniques and must be carefully managed to ensure that the theoretical benefits of precoding translate into real-world gains. There are basically two types of precoding

1. Basic precoding

2. Multi-cell precoding.

5. Pilot contamination

In massive multiple-input multiple-output (MIMO) systems, pilot contamination is a significant issue that arises during the process of channel estimation [21]. In these systems, the base station (BS) needs to estimate the channel state information (CSI) to optimize the transmission of signals to multiple users. This estimation is typically done using pilot signals, which are predefined sequences sent by users to help the BS estimate the channels.

However, when multiple users in different cells use the same or non-orthogonal pilot sequences, the BS may mistakenly attribute signals from users in neighboring cells to users in its own cell. This results in pilot contamination, which can severely degrade system performance, as the BS's channel estimates become corrupted, leading to poor beamforming, increased interference, and reduced data rates (**Figure 3**).

Types of pilot contamination:

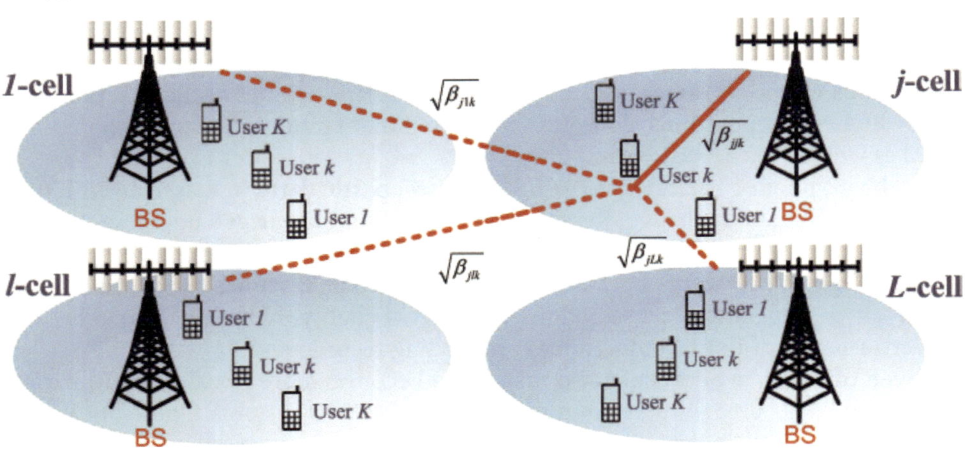

Figure 3.
Diagrams illustrating pilot contamination [21].

5.1 Inter-cell pilot contamination

This type occurs when users in different cells use the same pilot sequences. The BSs in adjacent cells cannot distinguish between the signals from their users and those from users in neighboring cells, leading to incorrect CSI estimation. This is the most common form of pilot contamination in massive MIMO systems.

5.2 Intra-cell pilot contamination

Although less common, intra-cell pilot contamination happens when users within the same cell use non-orthogonal pilot sequences. This can occur due to errors in pilot assignment or insufficient pilot sequence resources. The BS struggles to accurately estimate the CSI for each user, leading to interference within the cell.

5.3 Pilot contamination due to mobility

Users moving between cells can cause pilot contamination if the system does not update the pilot assignments quickly enough. As users enter new cells, they may use pilots that conflict with those in use in their new environment, causing CSI estimation errors.

5.4 Pilot contamination in multi-cell systems with reuse

In systems where pilot sequences are reused across multiple cells (e.g., in dense cellular networks), pilot contamination can occur even if cells are geographically distant. The reuse factor determines how frequently pilots are repeated across cells, with lower reuse factors leading to higher contamination risks.

6. Energy efficiency

In addition to enhancing spectral efficiency, massive MIMO technology significantly improves power efficiency as well. This section will first explore the power scaling law specific to massive MIMO systems and then delve into the trade-off between spectral and energy efficiency.

The power scaling law, essential for comprehending power efficiency in massive multi-user MIMO (MU-MIMO) systems, has been previously derived in studies like Ref. [4]. To contextualize this within a typical scenario, consider an uplink transmission in a single-cell system where there are K single-antenna users and a base station (BS) equipped with N antennas, with N being significantly larger than K. The channel vectors from different users are assumed to be uncorrelated, which is a common assumption in massive MIMO theory. In this setup, linear detectors—such as the matched filter (MF), zero-forcing (ZF), and minimum mean square error (MMSE) detectors—are sufficiently effective for signal processing. These detectors are not only simpler to implement but also offer near-optimal performance in massive MIMO systems, especially when the number of antennas at the BS becomes very large. The reason these linear detectors perform so well in massive MIMO systems lies in the power scaling law.

The power scaling law essentially states that as the number of antennas N at the BS increases, the total transmitted power required by the users can be scaled down proportionally without compromising the system's performance. Specifically, when N becomes very large, the transmitted power per user can be reduced significantly while still maintaining a reliable and high-quality communication link. This is because the large number of antennas allows the BS to accurately detect and separate the signals from each user, even in the presence of noise and interference.

For instance, with the MF detector and perfect channel state information (CSI) at the BS, the ergodic achievable uplink data rate for the k-th user increases as the number of BS antennas N approaches infinity. The ergodic rate reflects the average rate that can be achieved over a long period, accounting for the varying channel conditions. As N increases, the signal-to-noise ratio (SNR) improves due to the array gain provided by the large number of antennas, allowing for higher data rates even as the transmission power is reduced.

This power scaling law highlights a key advantage of massive MIMO systems: the ability to achieve high data rates with reduced power consumption. However, this brings us to the trade-off between spectral and energy efficiency. While spectral efficiency focuses on maximizing the data rate per unit of bandwidth, energy efficiency emphasizes minimizing the power consumption per unit of data transmitted. In massive MIMO systems, increasing the number of antennas improves both spectral and energy efficiencies to a point, but beyond that diminishing returns may occur.

The challenge, therefore, is to strike an optimal balance between these two objectives. By carefully managing the power allocation, number of antennas, and the choice of detectors, massive MIMO systems can achieve an optimal trade-off, ensuring both high spectral efficiency and excellent energy efficiency. This balance is crucial for the sustainable deployment of next-generation wireless networks, where both high performance and low energy consumption are critical requirements.

7. Conclusion

In this chapter, we have provided an extensive exploration of massive MIMO systems, examining them from multiple angles. The core idea behind massive MIMO is equipping a base station (BS) with a large array of antennas, which has the potential to greatly enhance both spectral and energy efficiency. By doing so, massive MIMO can support a higher number of users and deliver faster data rates while consuming less power, making it a promising technology for future wireless networks.

However, while the theoretical advantages of massive MIMO are clear, there are still several critical challenges that must be addressed to fully realize these benefits in practical applications. For instance, one of the key issues is channel correlation, which occurs when the signals received or transmitted by the antennas are not independent due to spatial proximity or other factors. This correlation can reduce the effectiveness of massive MIMO systems, diminishing the expected gains in capacity and efficiency.

Additionally, the hardware required to implement massive MIMO presents its own set of challenges. As the number of antennas increases, so does the complexity of the hardware needed to support them. This includes not only the physical antennas but also the associated signal processing units, power amplifiers, and other components. Moreover, hardware impairments, such as nonlinearities and phase noise, can further degrade the performance of massive MIMO systems, making it essential to develop robust techniques to mitigate these effects. Interference management is another crucial area of research. In dense networks where many users and base stations operate in close proximity, managing interference becomes increasingly complex. Massive

MIMO can help mitigate interference by focusing the signal energy more precisely in the direction of the intended users, but this requires advanced algorithms and coordination strategies to be fully effective.

Lastly, modulation techniques need to be adapted or developed to suit the unique characteristics of massive MIMO systems. The choice of modulation affects how efficiently the system can use the available spectrum and how resilient it is to noise and interference. Optimizing modulation schemes for massive MIMO will be key to achieving the desired improvements in spectral efficiency. In summary, while massive MIMO holds great promise for revolutionizing wireless communication by dramatically improving spectral and energy efficiency, realizing these benefits will require substantial further research. Addressing the challenges related to channel correlation, hardware limitations, interference management, and modulation techniques will be essential for the successful deployment of massive MIMO in future networks.

Author details

Bhawna Kalra[1,2], M.M. Sharma[1*] and Jaiverdhan[3]

1 Department of Electronics and Communication Engineering, Malaviya National Institute of Technology, Jaipur, India

2 Department of Electronics and Communication Engineering, Jaipur Engineering College and Research Centre, Jaipur, India

3 Department of Electronics Engineering, National Institute of Technology, Uttarakhand, India

*Address all correspondence to: mms.ece@mnit.ac.in

References

[1] Marzetta TL. Multi-cellular wireless with base stations employing unlimited numbers of antennas. In: UCSD Information Theory and Applications Workshop. 2010

[2] Marzetta TL. Noncooperative cellular wireless with unlimited numbers of base station antennas. IEEE Transactions on Wireless Communications. 2010;**9**: 3590-3600

[3] Larsson EG. Very Large MIMO Systems: Opportunities and Challenges. 2012. Available from: http://www.kth.se/polopolyfs/1.303070!/Menu/general/columncontent/attachment/LargeMIMO.pdf.

[4] Ngo HQ, Larsson EG, Marzetta TL. Energy and spectral efficiency of very large multiuser MIMO systems. IEEE Transactions on Communications. 2013; **61**:1436-1449

[5] Li GY, Xu Z-K, Xiong C, Yang C-Y, Zhang S-Q, Chen Y, et al. Energy-efficient wireless communications: Tutorial, survey, and open issues. IEEE Wireless Communications Magazine. 2011;**18**:28-35

[6] Xiong C, Li GY, Zhang S, Chen Y, Xu S. Energy- and spectral-efficiency tradeoff in downlink OFDMA networks. IEEE Transactions on Wireless Communications. 2011;**10**:3874-3886

[7] Rusek F, Persson D, Lau BK, Larsson EG, Marzetta TL, Edfors O, et al. Scaling up MIMO: Opportunities and challenges with very large arrays. IEEE Signal Processing Magazine. 2013;**30**: 40-46

[8] Larsson EG, Tufvesson F, Edfors O, Marzetta TL. Massive MIMO for next generation wireless systems [online]. IEEE Communications Magazine. 2013. Available from: http://arxiv.org/pdf/1304.6690v3.pdf

[9] Matthaiou M, MacKay MR, Smith PJ, Nossek JA. On the condition number distribution of complex wishart matrices. IEEE Transactions on Communications. 2010;**58**:1705-1717

[10] Vishwanath S, Jindal N, Goldsmith A. Duality, achievable rates, and sum-rate capacity of Gaussian MIMO broadcast channels. IEEE Transactions on Information Theory. 2003;**49**:2658-2668

[11] Stuber GL, Barry J, McLaughlin S, Li YG, Ingram MA, Pratt T. Broadband MIMO-OFDM wireless communications. Proceedings of the IEEE. 2004;**92**: 271-294

[12] Marzetta TL. How much training is required for multiuser MIMO? In: Fortieth Asilomar Conference on Signals, Systems and Computers (ACSSC), Pacific Grove, CA. 2006. pp. 359-363

[13] Nguyen S, Ghrayeb A. Compressive sensing-based channel estimation for massive multiuser MIMO systems. In: Proc. IEEE Wireless Communications and Networking Conference. (WCNC), Shanghai, China. 2013. pp. 2890-2895

[14] Dai L, Wang Z, Yang Z. Spectrally efficient time-frequency training OFDM for mobile large-scale MIMO systems. IEEE Journal on Selected Areas in Communications. 2013;**31**:251-263

[15] Hoydis J, ten Brink S, Debbah M. Massive MIMO in the UL/DL of cellular networks: How many antennas do we need? IEEE Journal on Selected Areas in Communications. 2013;**31**:160-171

[16] Liang Y-C, Pan GM, Bai ZD. Asymptotic performance of MMSE receivers for large systems using random matrix theory. IEEE Transactions on Information Theory. 2007;**53**(11): 4173-4190

[17] Ngo HQ, Matthaiou M, Duong TQ, Larsson EG. Uplink performance analysis of multiuser MU-SIMO systems with ZF receivers. IEEE Transactions on Vehicular Technology. 2013;**62**: 4471-4483

[18] Matthaiou M, Zhong C, McKay MR, Ratnarajah T. Sum rate analysis of ZF receivers in distributed MIMO systems. IEEE Journal on Selected Areas in Communications. 2013;**31**:180-191

[19] Liang Y-C, Sun S, Ho C. Block-iterative generalized decision feedback equalizers BI-GDFE for large MIMO systems: Algorithm design and asymptotic performance analysis. IEEE Transactions on Signal Processing. 2006; **54**:2035-2048

[20] Vardhan KV, Mohammed SK, Chockalingam A, Rajan BS. A low-complexity detector for large MIMO systems and multicarrier CDMA systems. IEEE Journal on Selected Areas in Communications. 2008;**26**:473-485

[21] Zuo J et al. Multicell multiuser massive MIMO transmission with downlink training and pilot contamination precoding. IEEE Transactions on Vehicular Technology. 2016;**65**:6301-6314

Chapter 4

Extension to Critical Analysis of Active Shielding Methods for Space Radiation Protection

Charles J. Baker and Steven J. Simske

Abstract

This work extends the foundational analysis presented in the original Lawrence Townsend paper on advanced active shielding techniques involving electromagnetic fields aimed at protecting astronauts from the hazardous space radiation of solar energetic particle (SEP) events and galactic cosmic rays (GCRs). In recent years, there has been ongoing interest in advanced active shielding techniques involving electromagnetic fields to safeguard astronauts from hazardous space radiation. Proposals have aimed to mitigate risks from solar energetic particle (SEP) events, galactic cosmic rays (GCRs), or a combination of these. However, many of these innovative designs are based on oversimplified or sometimes outdated understandings of radiation spectra, risk profiles, and technological constraints. Optimistic projections often overlook the significant challenges in transitioning these theoretical models into practical, deployable technologies. This paper provides an in-depth review and critical analysis of advanced active shielding methods for space radiation protection, focusing on electromagnetic field-based techniques. It evaluates various strategies proposed over the past decades to protect astronauts from solar energetic particle events and galactic cosmic rays, underscoring the technical and conceptual challenges in translating these theories into viable, deployable technologies. By highlighting the potential benefits and identifying technical drawbacks and analytical gaps, this revision aims to offer a balanced, updated perspective on the state of active radiation shielding technology for space missions.

Keywords: radiation, shielding, magnetic field, electromagnetic, lethal dose

1. Introduction

As humanity embarks on ambitious projects like manned Mars missions and lunar habitation, astronauts will confront unprecedented levels of space radiation, surpassing experiences from the Apollo missions, low-Earth orbit (LEO) endeavors, or stays on the International Space Station (ISS). The primary concerns in these deep space environments are sporadic Solar Energetic Particle (SEP) events and the persistent threat of Galactic Cosmic Rays (GCRs). SEPs pose a risk of acute radiation syndrome, alongside contributing to long-term health effects like cancer. Meanwhile, GCRs, especially the high-energy heavy ions known as HZE particles (High (H) atomic number (Z) and energy (E)), are predominantly associated with stochastic health risks, such as increased cancer rates and potentially unique health outcomes due to their penetrating and damaging nature [1]. To address these concerns comprehensively, it is crucial to integrate functional testing with biological models to observe the genetic impacts of radiation, which can complement the theoretical and engineering analyses of shielding technologies [2]. For decades, the potential of active shielding strategies, employing electromagnetic fields or plasma, to protect spacecraft, equipment, and crew from these intense radiation environments has been a subject of research. Active shielding holds the promise of mitigating health risks by significantly reducing or entirely deflecting harmful radiation, potentially offering a lighter and more cost-effective solution than traditional passive bulk material shielding. Various designs targeting protection from SEPs, GCRs, or both have been explored. However, many of these proposals are based on oversimplified or outdated understandings of radiation spectra and risks, coupled with overly optimistic assumptions about bridging significant technological gaps to realize these concepts. This paper aims to critically review and evaluate various active shielding techniques proposed over the years. It will detail each method's potential advantages and technical challenges and highlight any analytical gaps in assessing their shielding effectiveness. The core question is not merely about the feasibility of active shielding but rather whether the total mass of an active shielding system—including its components like coils, power sources, and support structures—is truly more mass-efficient than equivalent passive shielding in providing a specified level of radiation protection.

2. Active shielding methods

2.1 Technological advancements in material science and engineering

Recent advancements in material science and engineering, particularly in superconducting and nano-engineered materials, present promising avenues to revolutionize active shielding methods for space radiation protection. These innovative materials could dramatically reduce the weight and improve the efficiency and effectiveness of shielding systems, addressing some of the primary concerns highlighted in previous studies.

2.1.1 Superconducting materials

Advances in superconducting materials offer the potential to create more efficient magnetic shields. High-temperature superconductors, for example, can operate at relatively higher temperatures compared to traditional superconductors, reducing the need for complex and heavy cryogenic systems. This can significantly reduce the overall weight and increase the feasibility of magnetic shielding methods in space. Moreover, superconducting materials with higher critical magnetic fields and critical currents are being developed, enhancing the practical deployment of magnetic shields in the harsh space environment [3, 4].

I. Yttrium-barium-copper-oxide (YBCO): Noted for its high critical temperature, YBCO is identified as an efficient material for generating strong magnetic fields necessary for deflecting charged particles in space. Its advantages include the ability to operate at higher temperatures, which can significantly reduce the need for complex cooling systems in the spacecraft environment [5].

II. Bismuth strontium calcium copper oxide (BSCCO): This HTS material is another highlighted for its potential in active magnetic shielding applications. Similar to YBCO, BSCCO can generate strong magnetic fields but has different operational characteristics that might make it suitable for specific shielding configurations [5].

III. Magnesium diboride (MgB2): While not a cuprate superconductor like YBCO and BSCCO, MgB2 possesses a relatively high critical temperature and the potential for use in space radiation shielding applications. Its unique properties might offer advantages in terms of manufacturing or integration into magnetic shielding systems [5].

These materials can carry much greater current densities than common conductors without losing energy to electrical resistance. This remarkable property stems from their ability to enter a superconducting state, where electrical resistance drops to zero, allowing current to flow through without any energy loss due to heat. The capability to carry higher current densities is a result of the superconducting materials' unique electronic structure and the formation of Cooper pairs—pairs of electrons that move together through the material's lattice without scattering. This coherent movement allows superconductors to conduct electricity with perfect efficiency [6].

By carrying greater current densities without energy loss, these materials enable the creation of much stronger magnetic fields for a given amount of electrical current compared to traditional conductors. This efficiency is pivotal in reducing the size and weight of devices like magnets used in magnetic shielding systems. The stronger the magnetic field that a material can generate per unit of current, the less material is needed to achieve the desired magnetic field strength. Consequently, devices made with HTS materials can be smaller and lighter because they require less superconducting material to generate powerful magnetic fields. This reduction in size and weight is crucial for space applications, where every kilogram adds to the launch cost and complexity of space missions.

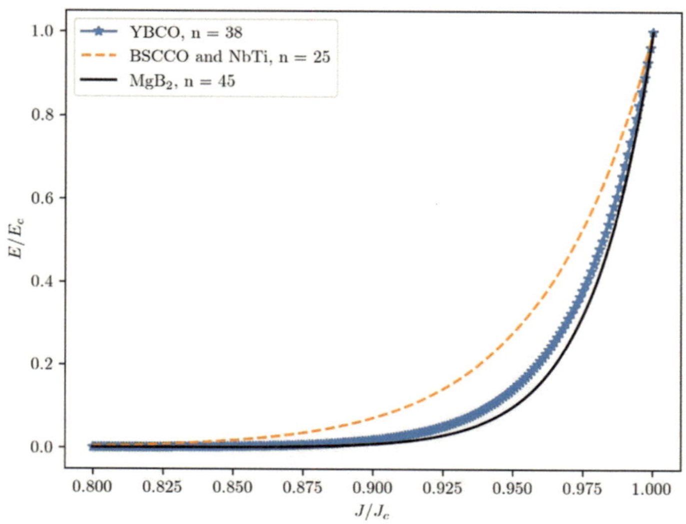

Figure 1.
Normalized power law for different superconducting materials with their typical n values. The electric field is E; the current density is J [6].

Figure 1 shows an example of the normalized E-J characteristic of stabilized type II superconductors from which resistivity can be inferred around the Jc. Up to Jc, the electric field is almost zero. As the current density approaches Jc, it increases sharply. Jcs are within the range of 108 A/m^2 for superconducting bulks and 1010 A/m^2 for superconducting tapes. This means the superconductor typically carries greater current densities than common conductors. This allows the reduction in device sizes and weight [6].

These advancements in high-temperature superconductors are crucial for the development of practical, efficient magnetic shielding systems for long-duration space missions. By leveraging the properties of HTS materials, researchers aim to create magnetic fields strong enough to protect astronauts from harmful space radiation while minimizing the energy and cooling requirements typically associated with superconducting magnets.

2.1.2 Nano-engineered materials

Nano-engineering allows for the design and synthesis of materials with tailored properties for specific applications, including radiation shielding. For instance, materials can be engineered to have enhanced radiation-absorbing properties or to be more effective at dispersing the energy of high-energy particles. This can lead to the development of lighter, more effective shielding materials that provide better protection against the wide spectrum of radiation encountered in space [7].

2.1.3 Advancements in cryogenics and magnet technology

Innovations in cryogenics contribute to the practicality of deploying superconducting materials in space by enhancing the efficiency and reliability of cooling systems. Additionally, advancements in magnet technology, such as the development of stronger and more lightweight magnets, can improve the design and implementation of magnetic shielding. These technologies are crucial for creating effective active shielding systems that can be used in long-duration space missions [5, 8].

.By incorporating these advancements, future spacecraft could be equipped with active shielding systems that are significantly more effective and practical. As material science and engineering continue to evolve, it is expected that new materials and technologies will emerge, further enhancing the prospects for effective active shielding in protecting astronauts from space radiation. However, it is critical to continue rigorous testing and validation of these materials and technologies in space-like environments to understand their behavior and ensure their effectiveness in mitigating radiation risks.

2.2 Plasma shielding

In the realm of plasma shielding, there have been significant breakthroughs in stabilizing plasma fields and reducing the power consumption necessary to generate and maintain these shields. The use of magnetic confinement has evolved, with novel geometries providing more stable plasma configurations that can be sustained with less energy. Breakthroughs in low-temperature plasma generation have also reduced the power requirements, making plasma shields more viable for spacecraft protection. Furthermore, research into plasma-material interactions has led to the development of materials that can withstand the harsh conditions of space and interact favorably with plasma, improving the durability and effectiveness of plasma shields. These

DOI: http://dx.doi.org/10.5772/intechopen.1007752

advancements address previous criticisms by offering energy-efficient and stable solutions, paving the way for their practical implementation in protecting spacecraft and crews from harmful space radiation.

Recent advancements in plasma stabilization and energy efficiency have focused on innovative geometric configurations of magnetic fields and plasma containment structures. Novel geometries, such as toroidal (doughnut-shaped) or spheromak (self-contained magnetic bubble) configurations, have been instrumental in creating more stable plasma fields. These shapes help to minimize the instabilities that typically arise in plasma, as they can more effectively contain the plasma and reduce the interaction with container walls, thereby lowering the energy losses. Additionally, advancements in magnetic confinement techniques, leveraging these geometries, have significantly reduced the power consumption needed to generate and maintain plasma shields by optimizing the magnetic field strength and distribution to keep the plasma stable with less external energy input [9].

As shown in **Figure 2**, toroidal configurations, like those used in tokamaks, confine plasma in a doughnut shape. This shape is inherently more stable due to the symmetry of magnetic fields that are smoothly distributed around the torus. This symmetry helps to reduce plasma instabilities and confine the plasma effectively. Spheromak configurations generate a self-contained magnetic bubble, where the plasma itself generates the magnetic fields necessary for its confinement, reducing the need for complex external field generators and leading to a more stable and self-sustaining plasma structure with reduced energy input. Both configurations aim to minimize areas where the plasma could come into contact with the containment vessel, thereby reducing energy losses and instabilities [10].

2.3 Electrostatic fields

Recent advancements in electrostatic field application for space radiation protection have focused on reducing the previously insurmountable power requirements and

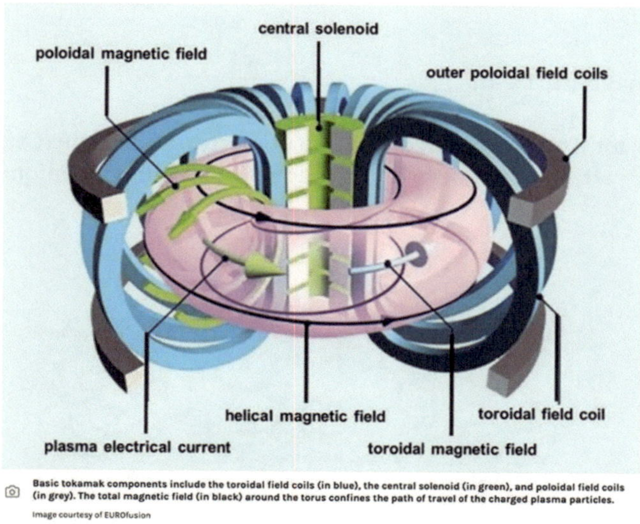

Figure 2.
Tokamak components [10].

achieving more stable configurations. Innovative materials and nano-technology have led to the development of lightweight, high-voltage electrostatic generators that are more feasible for space applications. These advancements have significantly lowered the energy requirements for maintaining electrostatic shields, making them a more practical option for long-duration space missions. Additionally, research into adaptive electrostatic field configurations has shown promise in dynamically adjusting field strengths and shapes to optimize shielding against varying radiation environments, thus enhancing the overall efficiency and stability of electrostatic shielding methods [11].

Nanotechnology contributes to the development of lightweight, high-voltage electrostatic generators through the use of nanomaterials and nanostructures, which offer superior electrical properties, high surface area-to-volume ratios, and enhanced conductivity. These nanoscale materials can be engineered to withstand high voltages while remaining compact and lightweight. Their unique properties facilitate the miniaturization of electrostatic generators, making them more efficient and suitable for space applications where size, weight, and power efficiency are critical. Additionally, nanotechnology allows for improvements in the durability and performance of these generators under harsh space conditions [12].

Examples of nanomaterials and nanostructures include carbon nanotubes (CNT), known for their exceptional strength and electrical conductivity; graphene, a single layer of carbon atoms with remarkable electrical and thermal properties; quantum dots, which have unique optical and electronic properties due to their quantum mechanical effects; and metallic nanoparticles, such as gold and silver, used for their electrical conductivity and antimicrobial properties. These materials are utilized in various applications, from electronics and energy storage to medical devices and environmental sensing, due to their unique properties at the nanoscale [7].

The axial electric current in the CNT induces a circumferential magnetic field within the CNT wall, as shown in **Figure 3**. Due to the magnetic field, the axial electric current subsequently experiences a Lorentz force. Effectively, this can be modeled as a pressure acting on the CNT wall. **Figure 3**, Pressure mechanism. (a) This illustration shows electromagnetic fields in a CNT relevant to the stiffening effect. An axial current (orange) is confined to the CNT walls and induces a circumferential magnetic field (blue).

2.4 Confined magnetic fields

Advances in confined magnetic fields, particularly in the context of space applications and spacecraft, focus on developing magnetic shielding techniques to protect

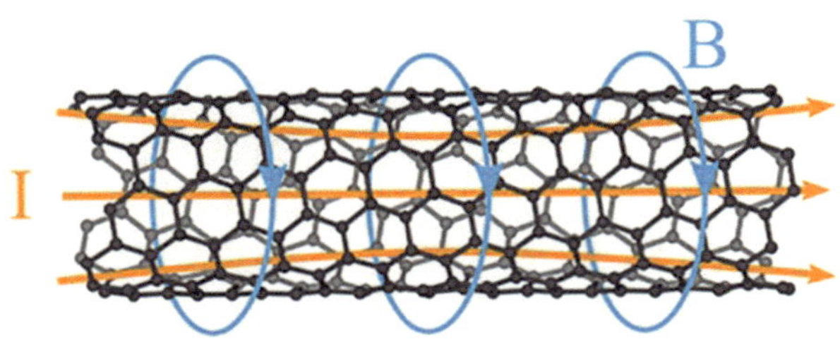

Figure 3.
Highly oriented direct-spun carbon nanotube textiles aligned by in situ radio-frequency fields with stiffening effect [7].

astronauts from harmful cosmic and solar radiation. The concept involves creating magnetic fields that mimic Earth's magnetosphere, providing a protective bubble around spacecraft or habitats. This area of research is crucial for long-duration space missions, including trips to Mars and extended stays on the Moon, where exposure to radiation poses a significant risk to human health.

2.4.1 Key advances and research areas

Magnetic field generation techniques:

Innovations in generating strong, yet confined magnetic fields are central to this research. This includes developing superconducting magnets that can create high-intensity fields with relatively low power consumption. The use of high-temperature superconductors has been a focal point, as they can operate at higher temperatures than traditional superconductors, potentially reducing the cooling requirements and overall system complexity.

Improved simulation tools and models have been developed to better understand and predict the behavior of confined magnetic fields in space. These tools help in designing more effective magnetic confinement systems by allowing researchers to simulate various configurations and their effectiveness in shielding against different types of radiation.

As shown in **Table 1**, these simulation tools and models have become increasingly sophisticated, leveraging advancements in computational physics, numerical methods, and computational hardware. These tools are pivotal in designing magnetic

Simulation tool/model	Purpose	Examples
Particle-in-cell (PIC) simulations	Simulate plasma dynamics and interactions between charged particles and electromagnetic fields. Model interactions between solar/cosmic charged particles and artificial magnetic fields around spacecraft.	OSIRIS, VORPAL
Magnetohydrodynamic (MHD) models	Simulate the behavior of ionized gases (plasmas) in the presence of magnetic fields. Study the global structure and stability of confined magnetic fields and their interaction with the solar wind.	BATS-R-US (Block-Adaptive Tree Solarwind Roe-Type Upwind Scheme)
Finite element method (FEM) models	Solve complex structural, material, and electromagnetic problems. Design magnetic coils and predict magnetic field distribution around spacecraft.	ANSYS Maxwell
Geant4	Simulate the passage of particles through matter. Evaluate the effectiveness of magnetic and material shielding strategies for spacecraft.	Geant4 simulations for modeling high-energy particle interactions with magnetic fields/materials
Hybrid simulations	Combine elements of PIC and MHD models for efficient plasma interaction simulations with magnetic fields over various scales. Study interactions between confined magnetic fields and solar wind or cosmic rays.	HYB simulation code

Table 1.
Simulation tools/models.

confinement systems, optimizing their configurations, and assessing their effectiveness in shielding spacecraft from various types of radiation.

Simulation tools/models provides a categorized overview of the tools and models commonly used in the study of magnetic fields and their interactions with space environments (**Table 1**). Each tool serves a distinct purpose and contributes uniquely to the overall design and evaluation process [13–15].

These simulation tools and models are instrumental in pushing the boundaries of what is possible with confined magnetic field technologies for space applications. By accurately modeling the complex interactions between magnetic fields and space radiation, researchers can devise more effective shielding strategies, enhancing the safety and feasibility of long-duration human space missions.

2.5 Unconfined magnetic fields

Current advances in unconfined magnetic fields, particularly for space applications, focus on extending magnetic protection to spacecraft and potential lunar or Martian bases by creating a dipole-like magnetic field around them. This approach aims to simulate the Earth's magnetosphere, providing a protective shield against harmful cosmic and solar radiation. The unconfined nature of these magnetic fields means they are not restricted to the immediate vicinity of the spacecraft or habitat but extend outward, creating a larger zone of protection. Recent advancements in this area include:

1. Magnetic field generation technologies advancements in generating large-scale magnetic fields have been crucial. Research has focused on developing efficient, high-temperature superconducting coils capable of producing strong magnetic fields with lower energy consumption. Innovations in materials science have enabled the creation of superconductors that work at higher temperatures, reducing the need for complex cooling systems [16].

2. Optimization of magnetic field configurations studies have been conducted to determine the optimal configurations for magnetic field generators to mimic a dipole field effectively. This includes the strategic placement of superconducting coils and the use of toroidal and poloidal field configurations to create a stable, Earth-like magnetosphere around a spacecraft or habitat. Computational models and simulations play a significant role in optimizing these configurations for maximum efficiency and protection [17].

3. Energy efficiency improvements significant efforts have been made to improve the energy efficiency of magnetic field generators. This is particularly important for space applications where power resources are limited. Advances in superconducting technology and magnetic field design have been aimed at reducing the power required to maintain a protective magnetic field over extended periods [18].

4. Radiation interaction models researchers have developed more sophisticated models to understand how different types of space radiation interact with unconfined magnetic fields. These models help in assessing the effectiveness of dipole-like magnetic fields in deflecting charged particles, including high-energy protons and heavy ions from solar wind and cosmic rays. Tools like Geant4 are used to simulate these interactions and optimize magnetic shield designs [14].

5. Experimental validation although much of the research is still in the theoretical and simulation stages, there have been proposals and initial tests aimed at experimentally validating the concepts of unconfined magnetic fields for space protection. These include ground-based experiments and the discussion of small-scale space missions to test magnetic field generators in the actual space environment [5].

6. Integration with spacecraft and habitat designs there's a growing focus on integrating magnetic field generation systems into the design of spacecraft, lunar bases, and Martian habitats. This involves not only the physical integration of superconducting coils but also considering the impact of the magnetic fields on onboard systems and living conditions [19].

2.5.1 Challenges and future directions

- Power consumption: Even with advances in superconducting materials and efficiency, maintaining a large-scale magnetic field in space requires significant power, posing a challenge for long-duration missions.

- Magnetic field strength and extent: Balancing the strength and extent of the magnetic field to provide adequate protection without interfering with spacecraft operations or requiring prohibitive amounts of energy. Though challenging, this can be done by optimizing the coil configurations using configurations like toroidal or Helmholtz coils that can focus the magnetic field effectively, enhancing its strength and reach where needed most. Utilizing cryogenic materials, though requiring cooling, offers high efficiency in magnetic field generation at lower energies by reducing thermal noise. Finally, magnetic flux pinning centers in superconductors help maintain a strong magnetic field with minimal energy input by stabilizing the magnetic flux lines [19].

- Material and system durability: Ensuring that the materials and systems used for generating unconfined magnetic fields can withstand the harsh conditions of space, including extreme temperatures, vacuum, and radiation, over long periods [20].

Unconfined magnetic field technologies represent a promising avenue for enhancing the safety of manned space missions by providing extensive radiation protection. Continued research and development in this area are critical for overcoming the existing challenges and making this technology a practical reality for future space exploration.

2.6 Shielding effectiveness criteria

Determining the effectiveness of spacecraft artificial electromagnetic shielding compared to passive, bulk material shields for radiation protection involves a comprehensive assessment that takes into account various criteria and the specific radiation environments encountered in space. The primary goal is to minimize the exposure of astronauts and sensitive equipment to harmful radiation, including solar particle events (SPEs), galactic cosmic rays (GCRs), and secondary radiation produced by interactions between primary radiation and shielding materials. The following are key criteria and considerations for evaluating shielding effectiveness:

2.6.1 Radiation types and energy Spectrum

Different types of radiation (e.g., protons, electrons, heavy ions) have varying levels of penetration and biological effects. An effective shield must address the specific types of radiation most prevalent in the spacecraft's operational environment.

In the spacecraft's operational environment, particularly in deep space beyond Earth's protective magnetosphere, astronauts and spacecraft are exposed to various types of space radiation. Each type of radiation has its own characteristics in terms of penetration ability, interaction with matter, and biological effects. The most prevalent types of space radiation include Galactic Cosmic Rays (GCRs), Solar Particle Events (SPEs), and trapped radiation belts near Earth. The following **Table 2** addresses each type [21].

Table 2 categorizes the different radiation sources and their effects, providing essential context for understanding the risks involved. Galactic Cosmic Rays (GCRs) consist primarily of high-energy protons (about 85 percent), with alpha particles making up about 14 percent, and a small fraction of heavier ions and high-Z elements known as High-Z and Energy (HZE) particles. GCRs are particularly dangerous due to their significant penetration abilities; HZE particles can easily penetrate spacecraft hulls and human tissue, leading to severe biological effects such as dense ionization, complex DNA damage, increased cancer risk, and neurological degeneration. This table emphasizes the severe impact GCRs can have, underscoring the need for robust shielding in deep space environments [2].

The energy distribution of incoming radiation, as outlined in the table, directly impacts shielding requirements. High-energy particles, particularly from GCRs and intense SPEs, pose a greater challenge for shielding due to their deeper penetration and potential to generate secondary radiation within the spacecraft. Understanding the specific characteristics of these radiation sources, as detailed in the table, is

Radiation source	Types of particles	Penetration levels	Biological effects
Galactic cosmic rays (GCRs)	High-energy protons (about 85%). Alpha particles (about 14%). Small fraction of heavier ions and high-Z elements (HZE particles)	Significant penetration abilities. High-Z and high-energy (HZE) particles can penetrate spacecraft hulls and human tissue	Dense ionization leading to complex DNA damage. Increased cancer risk. Neurological degeneration. Acute radiation syndromes. More severe effects due to
Solar particle events (SPEs)	Primarily protons with energies ranging from a few MeV to several GeV. Smaller quantities of electrons and heavy ions	Lower penetration abilities compared to GCRs. Intense solar events can produce high-energy protons significant enough to penetrate spacecraft	Acute radiation sickness (nausea, vomiting, fatigue). Increased long-term cancer risk. Serious threat to unprotected astronauts
Trapped radiation belts (Van Allen Belts)	Trapped protons and electrons of various energies, primarily from the solar wind	Varying penetration abilities. High-energy protons can pose a risk to spacecraft electronics and crew	Increased cancer risk. Acute radiation effects similar to SPEs. Exposure minimized by quickly passing through

Table 2.
Radiation types and effects.

crucial for designing effective protective measures for both astronauts and spacecraft systems. For instance, the variability in SPE particle flux underscores the importance of monitoring solar activity and adjusting shielding strategies accordingly.

The energy distribution of incoming radiation affects shielding requirements. High-energy particles pose a greater challenge for shielding due to their deeper penetration and potential to generate secondary radiation. The flux of Solar Particle Event (SPE) particles can vary significantly based on the intensity and type of solar event, such as solar flares or coronal mass ejections (CMEs). SPEs can range from relatively minor increases in particle flux to significant events where the particle flux can increase by several orders of magnitude, posing significant risks to astronauts and spacecraft systems. We can look at historical data from notable SPEs to provide an example of this variability.

One of the most significant recorded SPEs occurred in August 1972, between the Apollo 16 and Apollo 17 missions. During this event, the proton flux dramatically increased, measuring the number of protons passing through a square centimeter per second. The flux of protons with energies greater than 10 MeV reached levels of approximately 4.5×10^5 particles per square centimeter per second (particle/cm^2/s), which is orders of magnitude higher than typical background levels in space [22]. For context, the background flux of GCRs and lower-energy particles in space is much lower, typically well below 1 (particle/cm^2/s) for energies above 10 MeV. During quiet solar periods, the flux of particles from SPEs might be negligible or only slightly elevated above these background levels. It is important to note that the intensity and energy distribution of SPEs can vary widely:

Low-Intensity SPEs: These might only cause slight increases in radiation levels, which are potentially manageable with standard spacecraft shielding and pose minimal risk to crew health.

"Within the solar system, the GCR environment varies with the solar cycle in an inverse relationship, such that when the sun is undergoing intense activity, the GCR is at a lower intensity" [23]. **Figure 4** shows the total number of GCR element species per unit area at solar min and solar max.

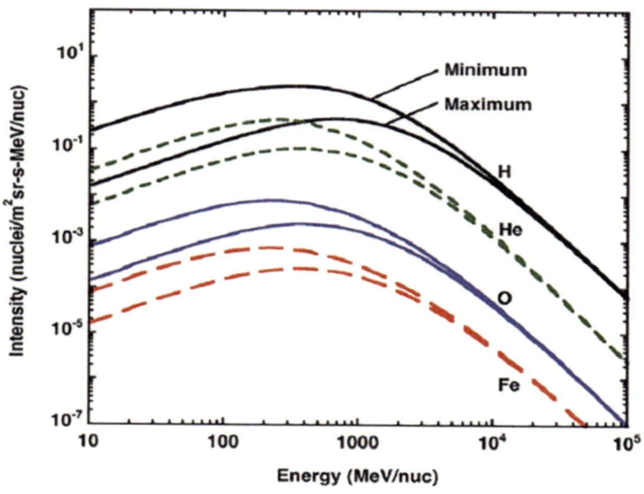

Figure 4.
Differential fluence of several GCR elemental species at solar minimum and solar maximum [23].

High-Intensity SPEs: Can result in extremely high fluxes of energetic particles, necessitating immediate protective measures, such as moving astronauts to shielded areas of the spacecraft and implementing operational changes to protect sensitive electronics. Due to solar activity's complex nature, predicting the flux of SPE particles is challenging. Satellite monitoring and space weather forecasting systems, such as those operated by NOAA's Space Weather Prediction Center (SWPC), provide real-time data and forecasts to help anticipate significant SPEs. SPEs are closely linked to the solar cycle, which lasts approximately 11 years and alternates between high and low solar activity periods, termed "solar maximum" and "solar minimum," respectively. During the solar maximum phase, the Sun exhibits more frequent and intense SPEs, characterized by the ejection of high-energy particles. Due to the heightened radiation levels, these events can significantly impact space weather, posing increased risks to spacecraft and astronauts. Understanding this cycle is crucial for predicting SPE occurrences and preparing protective measures for space missions. "Historical solar cycle relevance areas can be singled out (**Figure 5**). As can be seen from the graph, these areas have relatively few SPEs and are low intensity when compared with some of those SPEs that occurred during solar maximum conditions. The areas highlighted in green represent the range of sunspot numbers that are in accordance with the predicted sunspot values for April 2018 and are heading into solar minimum" [23].

The evaluation of spacecraft artificial shielding effectiveness for radiation protection is a multifaceted process that considers the nature of space radiation, the properties and configuration of the shielding material, the biological impact of radiation, the mass efficiency of the shield, and operational considerations. Advanced simulation tools and experimental validation play critical roles in optimizing shield designs to meet these criteria, ensuring crew and equipment safety on space missions. The following are characteristics to consider for shielding effectiveness:

The properties of shielding materials play a crucial role in determining their effectiveness against different types of radiation. Materials with high mass attenuation and absorption coefficients are especially effective at blocking high-Z (atomic number)

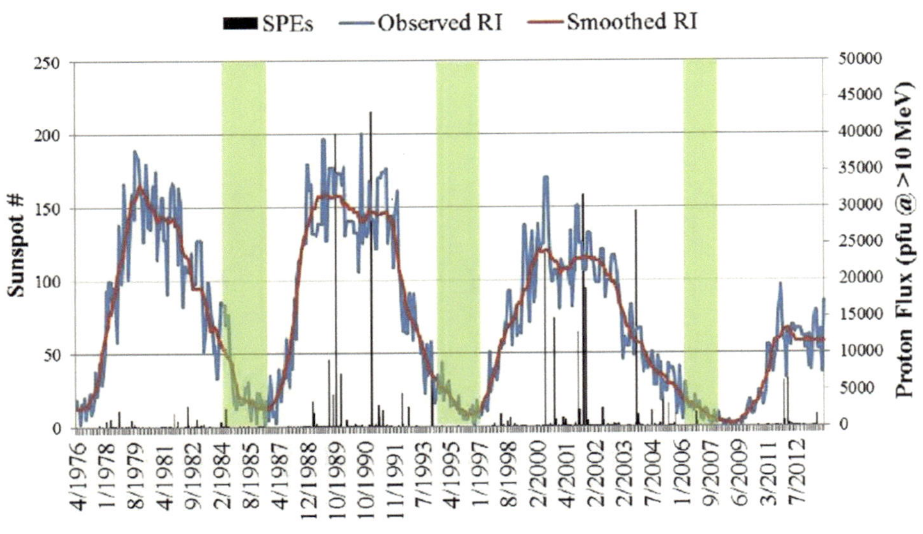

Figure 5.
SPE intensity (proton flux) plotted with the sunspot numbers from 1976 to 2012 [23].

particles. However, while some materials are adept at stopping primary particles, they can produce significant secondary radiation, including neutrons, which may be more biologically harmful than the primary radiation itself. The thickness and configuration of the shield are also vital factors. Often, there is an optimal thickness for a shield; beyond this point, additional material may not significantly increase protection and could even heighten risk due to the generation of secondary radiation. The spatial configuration and geometry of the shielding material can significantly affect its effectiveness, with certain arrangements offering superior protection against radiation from specific directions. In terms of biological effectiveness, the dose equivalent measure, which accounts for the absorbed dose and the biological impact of the radiation type, provides a more precise evaluation of the potential health risks from radiation exposure. Shielding effectiveness against acute doses from Solar Particle Events (SPEs) and chronic exposure to Galactic Cosmic Rays (GCRs) is particularly important. GCRs present a unique challenge due to their high energy and penetrating power, making effective shielding critical for long-term space missions [24].

The mass efficiency of the shielding system is another critical consideration, particularly for spacecraft design, as the mass impacts the cost of launching. Effective shields are those that optimize protection while minimizing mass. The effectiveness of electromagnetic shields is often compared with passive material shields to assess whether the increased complexity and power requirements of electromagnetic shields are justified by their enhanced protection or reduced mass. Operational considerations and reliability are also paramount. Electromagnetic shields, for instance, require power, which must be factored into the overall power budget and system design of a spacecraft. The reliability of the shielding system under long-duration space conditions, including its resilience to failures and the implementation of redundancy measures, is essential for ensuring continuous protection. Lastly, the shield's flexibility and adaptability to varying radiation conditions enhance its utility. The ability to adjust to changes in the radiation environment, such as those encountered during solar flares or in different orbital regions, is crucial for maintaining the shield's effectiveness over time [25].

Evaluating the effectiveness of spacecraft artificial electromagnetic shielding versus passive, bulk material shields for radiation protection necessitates a thorough assessment that encompasses a variety of criteria, particularly given the diverse radiation environments encountered in space. The overarching objective is to reduce exposure to harmful radiation, such as solar particle events (SPEs), galactic cosmic rays (GCRs), and secondary radiation generated from interactions between primary radiation and shielding materials. Key considerations include analyzing the types of radiation and their energy spectrum, as different radiations have distinct penetration levels and biological effects, with high-energy particles posing a significant challenge due to deeper penetration and potential secondary radiation production. The properties of shielding materials, including mass attenuation and absorption coefficients, are crucial, as materials effective against certain particles might produce harmful secondary radiation. The shield's thickness and configuration also play vital roles, with optimal thickness and spatial arrangements offering superior protection against directional radiation sources [5].

Biological effectiveness, gauged through measures like dose equivalent, which considers the type and biological impact of radiation, is paramount in assessing health risks. The shield's effectiveness against SPEs and GCRs, particularly the latter's high energy and penetration capability, is critical. Mass efficiency is another significant factor; effective shields are those that provide maximal protection with minimal mass, a consideration vital due to the cost implications of launching heavy payloads into space. Comparisons with passive shields evaluate whether the complexity and power requirements of electromagnetic shields are justified. Operational considerations, including power requirements and system reliability, affect overall spacecraft design and mission feasibility. Electromagnetic shields' need for power and their reliability, especially under long-duration space missions, along with the need for redundancy, are critical factors. Furthermore, the ability of the shield to adjust to varying radiation conditions enhances its effectiveness, offering a level of flexibility and adaptability that passive shields may lack [26].

3. EMF effects on RF communications and health

A spacecraft's artificial magnetic field system can cause RF (Radio Frequency) interference with communications primarily because the magnetic field can alter the path of the RF signals, potentially absorbing or deflecting them away from their intended path. This interference happens because electromagnetic waves, which include RF communications, interact with magnetic fields. Depending on the frequency of the RF signal and the strength and configuration of the magnetic field, this interaction can result in signal degradation or loss, impacting communication efficiency [27]. RF interference caused by a spacecraft's artificial magnetic field system can be mitigated through several methods:

1. Frequency management: Adjusting the frequencies used for communication to avoid those most affected by the magnetic field. Determining which frequencies are affected by the EMF (Electromagnetic Field) typically involves conducting electromagnetic compatibility (EMC) tests and simulations. These tests assess the interaction between the magnetic field and the spacecraft's communication systems across various frequencies. Simulation tools can model the electromagnetic environment of the spacecraft, predicting which frequencies are most susceptible to interference. This process helps identify the frequency bands that remain clear of interference or require adjustments to ensure reliable communication [27].

2. Shielding and filtering: Implementing additional shielding or filtering around communication equipment to protect against magnetic field interference. An example of implementing additional shielding or filtering around communication equipment to protect against magnetic field interference is the use of Faraday cages or enclosures around sensitive components. These enclosures are made from conductive materials and work by blocking external static and non-static electric fields. For filtering, ferrite beads or chokes can be placed on cables and wires to suppress high-frequency electromagnetic interference (EMI), allowing only the desired frequencies to pass through to the communication equipment. These methods effectively mitigate interference without significantly impacting the system's overall performance [28].

3. Spatial configuration: Designing the spacecraft's layout to physically separate the magnetic field generation system and the communication systems as much as possible.

4. Signal processing techniques: Using advanced signal processing techniques to detect and correct for the interference caused by the magnetic field. An example

of using advanced signal processing techniques to detect and correct for interference caused by the magnetic field involves the application of Adaptive Noise Canceling (ANC) and Digital Signal Processing (DSP) methods [28].

3.1 Adaptive noise canceling (ANC)

ANC works by using a reference signal that contains information about the noise or interference to be eliminated. In the context of magnetic field interference, the reference signal could be a measurement of the magnetic field's impact on the communication signal at a given moment. The ANC system then generates a signal that is the inverse of the noise and combines it with the original signal. This process effectively cancels out the noise from the magnetic field, allowing the underlying communication signal to be recovered more cleanly [29].

3.2 Digital signal processing (DSP)

DSP techniques can be specifically tailored to identify the characteristics of magnetic field interference in the frequency domain. By applying filters designed to remove or attenuate frequencies most affected by the magnetic field, DSP can enhance the quality of the received signal. Additionally, DSP can employ algorithms like Fast Fourier Transform (FFT) to analyze the signal in the frequency domain, identifying and isolating the parts of the signal distorted by the magnetic field interference. Both ANC and DSP techniques require sophisticated algorithms and computing power to process signals in real-time, making them highly effective for managing complex interference patterns and ensuring reliable communication in environments with electromagnetic interference, such as those encountered by spacecraft. These strategies can help maintain effective communication despite the challenges posed by artificial magnetic field systems [30].

3.3 EMF effects on human health

The EMF produced by a spacecraft's artificial magnetic field system can have various effects on the health of astronauts, though the extent and nature of these effects depend on the intensity, frequency, and exposure duration to the EMF. While the primary purpose of these systems is to protect crew members from harmful space radiation, ensuring that the EMF levels remain within safe exposure limits is crucial [31]. Potential health effects can include:

1. Neurological effects: High levels of EMF exposure have been associated with changes in nerve function, which could potentially lead to symptoms such as headaches, dizziness, and changes in sleep patterns.

2. Stress response: EMF exposure can lead to a stress response in the body, characterized by the release of stress hormones. Over time, this can affect immune system function and lead to various stress-related health issues.

3. Cognitive function: Research is ongoing into the effects of EMF on cognitive functions, including attention, memory, and spatial orientation. While conclusive evidence is still lacking, this is an area of concern for astronauts who rely heavily on these cognitive abilities.

It's important to note that spacecraft are designed with strict safety standards to minimize any potential negative health effects from EMFs. This includes careful design and placement of magnetic field generators, shielding of living and working areas, and adherence to international safety guidelines for EMF exposure [31]. An effective methodology for shielding astronauts from the EMF effects of the spacecraft's artificial magnetic field, while still allowing the system to protect them from radiation, involves a multi-layered approach:

1. Optimized shield design: Design the artificial magnetic field to be strong where it's needed for radiation protection but minimized in areas frequented by astronauts. This could involve strategic placement of field generators and using magnetic field shaping techniques.

2. EMF shielding materials: Use materials within the spacecraft's interior that can shield specific EMF frequencies without affecting the magnetic field's radiation protection capability. Materials like high-permeability metals or special composites can be used to line crew quarters or electronic equipment areas.

3. Distance and spatial configuration: EMF exposure can be reduced by increasing the distance between the magnetic field generators and the crew's living spaces. Designing the spacecraft to have a spatial configuration that places crew areas away from high EMF zones is crucial.

4. Active compensation systems: Implement active systems that monitor and neutralize excessive EMF levels in crewed areas. These systems can dynamically adjust to changing conditions, providing additional protection when needed.

5. Frequency management: Adjust the operating frequencies of the magnetic field generators and other onboard systems to avoid resonance with biological systems or critical communication frequencies.

6. Personal protection: For areas where EMF exposure might still be a concern, personal protective equipment designed to shield astronauts from EMF could be used. However, this is more of a last resort due to the impracticality for long-term use.

7. Continuous monitoring and control: Implement a comprehensive monitoring system to continuously assess EMF levels throughout the spacecraft. This will allow for real-time adjustments to the magnetic field strength or configuration to ensure astronaut safety.

This methodology combines engineering design, material science, and operational strategies to create a balanced approach that maximizes radiation protection while minimizing EMF exposure risks to astronauts [31].

In sum, determining the efficacy of artificial electromagnetic shielding in spacecraft involves a comprehensive evaluation of radiation types, shielding material properties, configuration, biological impact, mass efficiency, and operational considerations. This multifaceted approach ensures the safety of crew and equipment by optimizing shield designs through advanced simulation tools and experimental validation, addressing the complex challenges posed by space radiation.

Currently, no definitive career limits are established for deep space missions, such as those to Mars. Unlike missions in (LEO), where guidelines by the National Council on Radiation Protection and Measurements (NCRP) provide a framework for radiation exposure limits, the unique and prolonged exposure conditions of deep space missions necessitate a reconsideration of career limits. In this context, the concept of LD50, the lethal dose for 50 percent of a given population, can be instrumental in defining such limits. LD50, typically expressed in terms of radiation dose, reflects the acute and long-term effects of exposure on human health.

For deep space missions, career limits could be set based on an LD50 time factor, representing the cumulative time an astronaut can be exposed to space radiation before reaching a dose that corresponds to a 50 percent risk of lethality or serious health impairment over their career. This approach requires a deep understanding of the space radiation environment, individual susceptibility, and the effectiveness of protective measures [32].

As such, LD50 time becomes a critical factor in determining safe career durations for astronauts on missions beyond LEO, guiding the design of spacecraft shielding, mission planning, and crew selection. It is important that this metric be derived from a comprehensive model that considers the complexities of space radiation and human physiology, ensuring that the career limits set are both safe and feasible. **Figure 6** illustrates the concept of LD50. Just as with LEO missions, these limits would be expressed in terms of effective dose, taking into account the varying sensitivity of different body tissues to radiation and the specific geometry of the human body. This approach ensures a more accurate assessment of the potential risks and aids in the development of more effective protective strategies [32].

Based on current knowledge of radiation dosage in interplanetary space between Earth and Mars, a cumulative 1000 mSv (1 sievert) would probably cause fatal cancer many years later in five out of every 100 persons exposed to it.

Figure 6 is depicted using three different curves: Factor 1, Factor 2, and the Sum of Factors. Each curve represents the cumulative mortality rate due to different factors over time, plotted against the vertical axis that measures the percent of the population deceased and the horizontal axis representing time.

Factor 1 and Factor 2 show the mortality impacts of two separate factors, increasing the percentage of deaths over time at different rates. The Sum of Factors curve, which lies above the individual curves of Factor 1 and Factor 2, represents the combined effect of both factors and indicates a higher overall mortality rate.

The horizontal line drawn at the 50 percent mark corresponds to the LD50 point, where half of the population has been killed. The vertical lines labeled T1 and T2

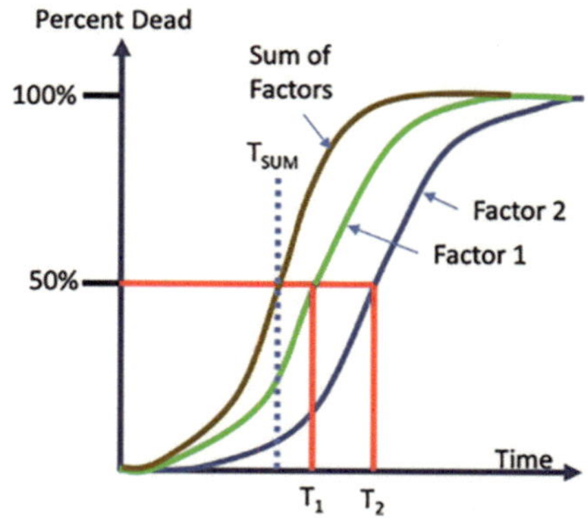

Figure 6.
Duration of radiation exposure vs. mortality percentage due to exposure [32, 33].

intersect the Factor 1 and Factor 2 curves at their respective LD50 points, where each factor alone results in 50 percent mortality. The dashed vertical line labeled T_sum intersects the Sum of Factors curve at the point where the combined effects of both factors reach the 50 percent mortality threshold.

Interestingly, T_sum is positioned to the left of T1 and T2, suggesting that the combined effects of the factors accelerate the time to reach LD50 compared to each factor alone. This indicates a synergistic effect, where the interaction between the two factors increases overall mortality more significantly than if their effects were simply additive. This graph thus effectively demonstrates how multiple factors can interact to affect the rate of mortality in a population [32].

4. Discussion

The exploration of advanced active shielding techniques presented in this paper reveals a complex interplay between theoretical promise and practical application challenges. These techniques, which are crucial for protecting astronauts from harmful space radiation, exhibit potential but also face significant hurdles that must be navigated as we push forward into deeper space exploration.

Protecting human life from hazardous space radiation is the ultimate goal of developing advanced shielding techniques. Traditional engineering tests and simulations, although essential, do not fully capture the biological implications of radiation exposure. By introducing functional testing with organisms that have well-understood genetic models and rapid reproduction rates, we can obtain direct empirical evidence of genetic damage and mutation rates induced by radiation. This data is crucial for assessing the biological efficacy of different shielding strategies [33, 34].

4.1 Technological viability and challenges

The viability of electromagnetic and plasma shields in actual space conditions remains uncertain, with significant challenges related to mass efficiency, thermal management, and system durability under cosmic radiation. While theoretical models are optimistic, real-world application requires overcoming substantial technological gaps. One such gap is the lack of empirical biological data to validate these models. Functional testing using organisms like *Escherichia coli* and *Saccharomyces cerevisiae* can provide critical insights into the biological efficacy of different shielding methods, highlighting genetic damage and mutation rates induced by radiation under various shielding conditions [32, 34]. Another technological gap example is the integration of superconducting materials in shielding applications suggests substantial benefits in magnetic field generation, but the real-world application is still limited by issues such as cryogenic cooling requirements and the mechanical strength of these materials in the space environment.

4.2 Comparison with traditional methods

Compared to traditional passive shielding methods, the advanced active shielding techniques discussed might offer reduced mass and enhanced flexibility. However, their complexity and power requirements pose practical deployment challenges. This comparison highlights the necessity of developing a hybrid shielding strategy that incorporates both passive and active elements to optimize overall mass, efficiency, and protective capability.

4.3 Adaptability to space environment

The dynamic nature of space radiation environments requires adaptable shielding solutions that can respond to varying intensities and types of radiation. The potential for real-time modulation of shielding properties based on environmental data represents a promising area for future research. This adaptability could significantly enhance mission safety and operational flexibility, reducing the risk to human life and sensitive equipment. When comparing different shielding methodologies, it is essential to include both SPE and GCR contributions in the space radiation environment assessment. In addition, shielding requirements should be based on realistic radiation levels, grounded in the actual risk to humans, rather than arbitrary limits. Finally, consider the entire mass of the shielding system, including the inherent protection provided by spacecraft structures and internal components. Astronauts will rely on more than just a magnetic field for safety and habitation.

4.4 Future research directions

Moving forward, it is imperative that research continues in the experimental validation of these shielding techniques under simulated and actual space conditions. Developing new materials tailored for the space environment, along with advanced computational models to simulate their behavior, will be crucial. Additionally, exploring the synergy between different shielding types and their integration into spacecraft design requires more focused investigation.

4.4.1 Integration of biological testing for shielding validation

The exploration of advanced active shielding techniques discussed in this paper primarily focuses on theoretical and engineering aspects. While these discussions are foundational, the integration of biological testing could significantly enhance the validation and refinement of these technologies. Specifically, the use of fast-turnover organisms like bacteria (*Escherichia coli*) and yeast (*Saccharomyces cerevisiae*) offers a unique opportunity to directly observe the biological impacts of radiation under various shielding conditions. This approach not only underscores the technological innovations but also ensures that these advancements translate into tangible safety benefits for space missions [34, 35].

To achieve this, a space-simulated environment can be used to expose both a control group (with no shielding) and various test groups (each shielded by different active shielding techniques such as electromagnetic fields, plasma shielding, etc.) to space radiation conditions simulated by solar energetic particles and galactic cosmic rays. Organisms can be divided into a control group receiving no radiation and several experimental groups each subjected to radiation through different shielding configurations. This setup will allow for the evaluation of the protective efficacy of each shielding method against a no-shielding baseline. Radiation exposure experiments can be conducted using sources that simulate solar energetic particles and galactic cosmic rays, with specialized radiation sources and dosimetry equipment to accurately mimic the space radiation environment. Varying levels of radiation will be applied to simulate different intensities of space radiation, including extreme solar events [36].

Samples from each group of organisms can be collected both before and after radiation exposure. This dual-timepoint sampling is critical for identifying mutations that occur due to the radiation exposure. Mutation rates can be calculated from the sequenced data to quantify the genetic damage inflicted by radiation under different shielding conditions. High-throughput DNA sequencing technologies such as Next-Generation Sequencing (NGS) can be employed to detect and analyze genetic changes with high sensitivity. Bioinformatics tools would be used to process sequencing data, identify mutations, and quantify mutation rates. Statistical methods, like ANOVA or t-test, would be applied to compare mutation frequencies between control and shielded groups to assess the genetic impact of each shielding strategy [37].

This approach offers multiple benefits. Direct observation of genetic impact measures the impact of radiation on genetic material, providing a clear picture of the biological effectiveness of shielding technologies. The fast turnover rates of chosen organisms allow for quick collection of data, enabling timely improvements and adaptations in shielding strategies. Additionally, using bacteria and yeast is cost-effective compared to higher organisms, making large-scale or repeated experiments feasible. This functional testing framework would not only validate the effectiveness of the shielding in a biological context but also contribute to a safer and more scientifically grounded approach to protecting astronauts during deep-space missions [33, 34].

4.5 Limitations

One of the primary limitations of this study is the reliance on theoretical models and simulation data, which may not fully capture the complexities of the space environment. Future studies should aim to incorporate more empirical data from space missions to validate and refine the proposed models.

5. Conclusions

This paper has systematically reviewed and provided a critical analysis of various active shielding strategies explored over recent decades for protecting astronauts from hazardous space radiation. It has presented the potential benefits and technical challenges associated with each type of active shield, including electrostatic fields, plasma shields, confined magnetic fields, and unconfined magnetic fields, while also highlighting areas where the analyses of shielding effectiveness may fall short. A significant addition to these analyses is the incorporation of functional testing with biological models. This approach not only validates the engineering solutions but also ensures that they translate into real-world biological protection, ultimately enhancing the safety and effectiveness of shielding technologies for long-duration human space missions. After systematically reviewing various active shielding approaches for mitigating space radiation risks, it becomes evident that while active shielding holds promise, its real-world applicability hinges on overcoming significant technological and mass-efficiency challenges. The comparison with passive shielding underscores the necessity of a comprehensive understanding of the radiation environment and the integration of advanced material science to meet the safety requirements of future deep space explorations. This paper emphasizes the need for continued research, development, and experimental validation to refine active shielding technologies, ensuring the safety and viability of prolonged human presence in space.

The optimism often seen in theoretical models of active shielding needs a tempered perspective that considers practical deployment challenges. This includes the technological gaps that remain in translating these models into deployable systems capable of withstanding the harsh conditions of space. For instance, while superconducting materials and plasma-based shields show promise, their effective implementation in space remains hampered by significant hurdles related to mass efficiency, thermal management, and the durability of the components under cosmic radiation exposure.

The synergy between various shielding types, such as the combination of magnetic and plasma fields, could potentially enhance protective outcomes but also introduces complexity in system design and energy management. It is imperative that future research continues to explore these combinations, guided by both experimental data and advanced simulations, to ascertain the most effective configurations.

Furthermore, our analysis underscores the importance of adaptability in shielding solutions. Space missions encounter dynamic radiation environments that demand shields capable of adjusting to varying intensity and types of radiation. The ability of a shielding system to modulate its response based on real-time environmental data could significantly enhance crew safety and mission success.

While the path forward is fraught with challenges, the continued evolution of material science, combined with innovative engineering solutions and rigorous testing, will be key to developing effective radiation shielding technologies. As humanity ventures further into space, the importance of developing robust shielding strategies to protect astronauts from the severe risks posed by solar energetic particles (SEPs) and galactic cosmic rays (GCRs) cannot be overstated. Our discussion extends the foundational ideas presented in Townsend's original work, highlighting recent advancements and identifying critical gaps that need to be addressed to make these technologies viable.

Conflicts of interest

Charles Baker was/is employed by the NASA/Jet Propulsion Laboratory during the Colorado State University graduate study when this work was completed and they sponsored the graduate expenses, but JPL had no role in the design of the study; in the collection, analyses, or interpretation of data; in the writing of the manuscript; or in the decision to publish the results.

Abbreviations

MgB2	magnesium diboride
YBCO	yttrium-barium-copper-oxide
BSCCO	bismuth strontium calcium copper
SEP	radiation of solar energetic particle
GCR	galactic cosmic rays
LEO	low earth orbit
ISS	international space station
HZE	(high (H) atomic number (Z) and energy (E) particles
HTS	high temperature superconductors
CNT	carbon nanotubes
PIC	particle in cell
MHD	magnetohydrodynamic

SWPC	space weather prediction center
EMF	electromagnetic field
EMC	electromagnetic compatibility
EMI	electromagnetic interference
RF	radio frequency
ANC	adaptive noise canceling
DSP	digital signal processing
FFT	fast fourier transform
NCRP	national council on radiation protection and measurements
LD50	lethal dose for 50 percent

Author details

Charles J. Baker* and Steven J. Simske
Colorado State University, Fort Collins, CO, USA

*Address all correspondence to: charles.j.baker@colostate.edu

References

[1] Townsend L. Critical analysis of active shielding methods of space radiation protection. In: Proceedings of the 2005 IEEE Aerospace Conference. Big Sky, MT, USA: IEEE; 2005

[2] Hu S. Solar Particle Events and Radiation Exposure in Space. Houston, TX: KBRwyle; 2017. Available from: https://three.jsc.nasa.gov/articles/Hu-SPEs.pdf

[3] Hull JR, Murakami M. Applications of bulk high-temperature superconductors. Proceedings of the IEEE. 2004;**92**(10):1705-1707

[4] Chandler D. MIT News, Tests Show High-Temperature Superconducting Magnets are Ready for Fusion. 2024. Available from: https://news.mit.edu/2024/tests-show-high-temperature-superconducting-magnets-fusion-ready-0304

[5] Ferrone K et al. A review of magnetic shielding technology for space radiation. Radiation. 2023;**3**:46-57

[6] Santos B et al. A review of technology readiness levels for superconducting electric machinery. Energies. 2023;**16**:5955. DOI: 10.3390/en16165955

[7] Issman L et al. Highly oriented direct-spun carbon nanotube textiles aligned by In situ radio-frequency fields. ACS Nano. 2022;**16**(6):9583-9597

[8] Green MA. Cryogenic techniques for large superconducting magnets in space. Cryogenics. 1989;**29**:487-489. NASA Technical Reports Server (NTRS). ISSN: 0011-2275

[9] Murray CA et al. Basic research needs for microelectronics. In: Report of the Office of Science Workshop on Basic Research Needs for Microelectronics. Oak Ridge, TN: U.S. Department of Energy Office of Scientific and Technical Information; October 23 – 25, 2018. Available from: https://www.osti.gov/servlets/purl/1616249

[10] Lanctot M. DOE Explains... Tokamaks. National Spherical Tokamak Experiment Upgrade. Washington, DC: Office of Science U.S. Department of Energy; 2022. Available from: https://www.energy.gov/science/doe-explainstokamaks

[11] Adams JH, Parnell TA, Winglee RM. Revolutionary Concepts of Radiation Shielding for Human Exploration of Space. Alabama: National Aeronautics and Space Administration, Marshall Space Flight Center, MSFC; 2005

[12] Shiza M, Khalid M, Yasir W. Nanotechnology: A revolution in modern industry. Molecules. 2023;**28**(2):661. DOI: 10.3390/molecules28020661

[13] Bettini P et al. Fast Fourier transform-volume integral: A smart approach for the electromagnetic design of complex systems in large fusion devices. Plasma Physics and Controlled Fusion. 2021;**63**:025010. DOI: 10.1088/1361-6587/abce8f

[14] Allison J et al. Recent Developments in Geant4. Nuclear Instruments and Methods in Physics Research Section A: Accelerators, Spectrometers, Detectors and Associated Equipment. 2016;**835**(2016):186-225

[15] Pritchett PL. Particle-in-cell simulation of plasmas—A tutorial. In: Space Plasma Simulation. Lecture Notes in Physics. Vol. 615.

Berlin, Heidelberg: Springer; 2003. DOI: 10.1007/3-540-36530-3-1

[16] Betancourt LR et al. High-Temperature Superconductors as a Key Enabling Technology for Space Missions. Las Vegas, NV: NeutronStar Systems UG; 2021. DOI: 10.2514/6.2021-4110

[17] Kashikhin VS et al. Torus CLAS12-Superconducting Magnet Quench Analysis. United States: N. p; 2014. DOI: 10.1109/TASC.2014.2299531

[18] Al Zaman A et al. Study on superconducting magnetic shield for the manned long termed space voyages. The Egyptian Journal of Remote Sensing and Space Science. 2021;**24**(2):203-210

[19] Westover S et al. Magnet architectures and active radiation shielding study (MAARSS). In: Final Report for NASA Innovative Advanced Concepts Phase II. NASA Broad Agency Announcement NNH12ZUA004N. Houston, TX: NASA Johnson Space Center; 2014. Available from: https://www.nasa.gov/wpcontent/uploads/2017/07niac2012 phaseiiwestoverradiationprotectionand architecturetagged.pdf

[20] Tripathi R. Meeting the grand challenge of protecting Astronaut's health: Electrostatic active space radiation shielding for deep space missions. In: NASA NIAC 2011 Supported Study (Final Report). Hampton, VA: NASA Langley Research Center; 2011. Available from: https://www.nasa.gov/wp-content/uploads/2017/07/niac2011phasei tripathi electrostaticactivespace tagged.pdf

[21] Types of Radiation in Space. NP-2014-03-001-JSC. 2015. Available from: https://www.nasa.gov/wp-content/uploads/2015/11/np-2014-03-001-jsc-orionradiation handout.pdf

[22] Coffey HE, editor. Collected data reports on august 1972 solar-terrestrial events. In: Report UAG-28. Boulder, CO: NOAA; 1973

[23] Rojdev K. Radiation Analysis of the Total Ionizing Dose for the Resource Prospector Mission. Houston, TX: NASA Technical Reports Server, American Institute of Aeronautics and Astronautics; Mar 2015

[24] Chang Q et al. Radiation shielding polymer composites: Ray-interaction mechanism, structural design, manufacture and biomedical applications. Materials and Design. 2023;**233**:112253. DOI: 10.1016/j.matdes.2023.112253

[25] Wilson JW et al. Shielding Strategies for Human Space Exploration. NASA Conference Publication 3360. Houston, TX: NASA Johnson Space Center; Dec 1997

[26] Warden D, Bayazitoglu Y. New comparative metric for evaluating spacecraft radiation shielding. Aerospace Research Central. Rice University, Houston, Texas. 2018;**35**(3):652-655

[27] Ma MT. EMC standards and regulations: A brief review. In: NISTIR 3989. Electromagnetic Fields Division. Electronics and Electrical Engineering Laboratory, Technology Administration, National Institute of Standards and Technology, Boulder, CO; 1992. Available from: www.govinfo.gov

[28] Sufyan M. EMI Shielding: Protecting Electronic Devices in a Noisy World. Wevolver, Overhoeksplein, Amsterdam, North Holland: Wevolver; 2024. Available from: https://www.wevolver.com/article/emi-shielding-protecting-electronic-devices-in-a-noisy-world

[29] Kaunitz J. Adaptive Filtering of Broadband Signals as Applied to

Noise Cancelling. Stanford Electronics Laboratories, Stanford University, Stanford, California; 1972, Rep. SU-SEL-3-038, August 1972 (Ph.D. dissertation) OCLC 15201972

[30] Li B et al. Electro-magnetic analysis of high-frequency digital signal processors. Springerplus. 2016;**5**:1313. DOI: 10.1186/s40064-016-2999-2

[31] Bailey WH. Health effects relevant to the setting of EMF exposure limits. Health Physics. 2002;**83**(3):376-386

[32] Simske S. The LD50 for Different Aspects of Spaceflight, ENGR 533, Spaceflight and Biological Systems, Spring 2023, Colorado State University, Walter Scott, Jr. College of Engineering, Fort Collins, CO; 2023

[33] Simske S. ENGR 533, Module 10, Spaceflight and Biological Systems (Space Life Sciences), Colorado State University, Walter Scott, Jr. College of Engineering, Fort Collins, CO; 2023

[34] National Research Council (US) Committee on the Biological Effects of Ionizing Radiation (BEIR V). Health Effects of Exposure to Low Levels of Ionizing Radiation. Beir V. Washington (DC): National Academies Press; 1990, 1990. 2, Genetic Effects of Radiation. Available from: https://www.ncbi.nlm.nih.gov/books/NBK218706/

[35] Al-Khaza'leh KA, Al-Fawwaz AT. The effect of static magnetic field on E. Coli, S. Aureus and B. Subtilis viability. Journal of Natural Sciences Research. 2015;**5**(24):153-156. Available from: www.iiste.org. ISSN: 2224-3186 (Paper) ISSN: 2225-0921 (Online)

[36] Vral A, Fenech M, Thierens H. The micronucleus assay as a biological dosimeter of in vivo ionising radiation exposure. Mutagenesis. 2011;**26**(1):11-17. DOI: 10.1093/mutage/geq078

[37] Youk J, Kwon YJ. Quantitative and qualitative mutational impact of ionizing radiation on normal cells. Cell Genomics. 2024;**4**(2):100499. DOI: 10.1016/j.xgen.2024.100499

Section 3

Massive MIMO for Communication and Cellular Networks

Chapter 5

The Family of LML Detectors and the Family of LAS Detectors for Massive MIMO Communications

Yi Sun

Abstract

The family of local maximum likelihood (LML) detectors, including the global maximum likelihood (GML) detector, and the family of likelihood ascent search (LAS) detectors are akin to each other and possess common properties significant in both theory and practical multi-input multi-output (MIMO) communications. It is proved that a large MIMO channel possesses the LML characteristic, implying and predicting that a local search detector with likelihood ascent, like a wide-sense sequential LAS (WSLAS) detector, can approach the GML detection. By the replica method, the bit error rate (BER) of an LML detector in the large MIMO channel is obtained. The BER indicates that, in the high signal-to-noise ratio (SNR) regime, both the LML and GML detectors achieve the AWGN channel performance when the channel load is as high as up to 1.5086 bits/dimension with an equal-energy distribution, and the channel load can be higher with an unequal-energy distribution. The analytical result is verified by simulation in the equal-energy distribution, showing that the sequential LAS (SLAS) detector, a linear-complexity LML detector, can approach the BER of the NP-hard GML detector. The LML and LAS detectors in the two families are successfully applied to symbol detection in massive antenna MIMO communications and demonstrate the performance near the GML detection. This book chapter reviews the LML and LAS detectors in a unified framework. The focus is on their formulation, relationships, properties, and GML performance in BER and spectral efficiency in large MIMO channels.

Keywords: multiple-input multiple-output, global maximum likelihood, local maximum likelihood, likelihood ascent search, next-generation wireless network

1. Introduction

The family of likelihood ascent search (LAS) detectors originated from the modified Hopfield neural network (MHNN) [1–8]. When the original Hopfield neural network (HNN) [9] is applied to image restoration, a least squares problem with integer support, the error metric change is nonmonotonic – a phenomenon of instability [10]. To guarantee stability, the sequential MHNN [1] is proposed to update neurons one by one and check the error metric decrease in each step. The MHNN [2] with the parallel, partial parallel, and sequential updating modes is proposed for binary image restoration, which sets up the corresponding thresholds to ensure stability. The MHNN with the eliminating highest error (EHE) criterion is proposed in Refs. [3, 4] that updates one or multiple neurons with the largest error gradients in each step to converge to a better solution in fewer steps. A generalized updating rule (GUR) is proposed in Refs. [5, 6], by which an arbitrary set of neurons can be updated in each step while the error metric is guaranteed to decrease monotonically. In Ref. [7], several algorithms with particular sequences of updating modes are derived from the GUR and applied to gray image restoration and reconstruction. The EHE criterion and the fastest metric descent (FMD) criterion are used to design the sequence of updating modes so that the algorithms converge in fewer steps to better solutions with smaller errors and fewer strips in simulation. Analytical results [8] confirm that the EHE and FMD-based algorithms can achieve higher correct transition probabilities.

When applied to code division multiaccess (CDMA) communications, these algorithms under the GUR turn out to perform likelihood ascent searches and form the family of LAS detectors [11–14]. Given any sequence of sets of updating candidates, the LAS detector searches out a sequence of bit vectors with monotonic likelihood ascent and, therefore, monotonically decreases the error probability of the bit vectors. In the LAS family, the wide-sense sequential LAS (WSLAS) detectors are particularly interesting and converge to a local maximum likelihood (LML) vector with neighborhood size one, the best detectors in the LAS family. The EHE and FMD criteria can be applied to enable the WSLAS detector to converge in fewer steps to a better solution [11]. By computationally efficiently updating the likelihood gradient in each step, all LAS detectors have a per-bit complexity linear in the number of transmitted bits. The concept of the LML detector is extended to any neighborhood size and forms the family of LML-J detectors with any neighborhood size $J \geq 1$ [15–17]. The LML-J detector can be realized by a likelihood ascent search in a neighborhood of size J and the computation is reduced to the minimum by efficiently updating the likelihood gradient. The per-bit computational complexity of the LML-J detector is in the order of $O\left(\sum_{i=1}^{J}\binom{K}{i}\right)$ with K transmitted bits. The WSLAS/LML-1 detectors belong to both the LAS and LML families. The bit error rate (BER) and asymptotic multi-bit efficiency (AME) of LAS detectors are analyzed in [18]. It is obtained that the LAS BER and AME are comparable to those of the global maximum likelihood (GML) detector [19] and demonstrate the superiority of LAS detectors over other detectors.

In large random spreading (LRS) CDMA channels, a random spectral-temporal spreading sequence spreads a bit, and both the number of bits and sequence dimension tend to infinity with their ratio kept constant. It is observed for the first time in simulation [20] that in a random spreading CDMA system, as the number of bits increases, the sequential LAS (SLAS) detector (a special WSLAS detector) monotonically decreases BER. When the number of bits reaches 200, the SLAS detector already approaches the single-bit BER under a channel load of 0.5 bits/s/Hz and signal-to-noise ratio (SNR) of 8 dB with an equal-energy distribution. It is essential that the LRS CDMA channels are proven to possess the "LML characteristic" [21, 22]; that is, the channel output retains the local topology of the input set of bit vectors. Therefore, "to achieve the GML detection one would, without the exhaustive search over the entire set, perform only an LML-1 detection." It is shown that the WSLAS/LML-1 detector achieves unit AME with channel load $< 1/2 - 1/(4 \ln 2)$ bits/dimension and, therefore, achieves the single-bit BER in the high SNR regime. In simulations, for an equal-power distribution, the WSLAS/LML-1 detector with a linear per-bit complexity approaches the NP-hard GML BER in all SNR with channel load as high as 1.05 bits/dimension [21, 22]. By employing sparse spreading sequences where most chips are zero, the WSLAS/LML-1 detectors need only about 16 nonzero chips to achieve the same BER performance with fully spread sequences [23, 24]. The analytical results in [21, 22] are extended to the general large MIMO channel in [25]. The BER of the WSLAS and LML-J detectors with fixed J is obtained by using the powerful replica method. It is shown that the WSLAS/LML-1 detector consistently achieves one solution of the GML BER in all channel conditions. In a wide practical range of channel load and SNR, the WSLAS/LML-1 detector performs identically to the GML detector and achieves spectral efficiency as high as 1.5086 bits/dimension and the single-bit BER bound in the high SNR regime in an equal-energy distribution.

When the LAS detector is applied to massive antenna MIMO systems in Ref. [26], it demonstrates excellent performance in low BER and linear per-bit complexity. The results triggered the extensive study of the family of LAS detectors and the development of low-complexity near-optimum detectors with local search [27–58]. Various detectors are used as the initial detector of the LAS detector [26, 28–35]. The reactive tabu search is applied to avoid previously visited vectors in the future search [36–42]. The set of candidate vectors in each step is appropriately selected and reduced to improve BER performance and reduce computational complexity [43–46]. The LAS detector starts with multiple initial vectors and searches through several sequences of vectors, and then the best output vector is chosen [47, 48]. The LAS detector and the LML-J detector for $J = 2, 3$ are cascaded to improve the BER performance at the cost of increased computational complexity [28, 49]. The performance of the LAS detector is investigated when channel state information is imperfect [50–52]. The LAS detector is applied to joint symbol detection and channel estimation [53], decoding [54], and antenna selection [55]. The LAS detector invoked specific hardware design [56] and implementation on an FPGA chip [35]. The LAS detector is combined with the convolutional neural network [57] and deep learning [58]. The key feature of the family of LAS detectors is to search a sequence of bit vectors with a monotonic likelihood ascent and converge to a fixed point in a finite number of search steps. The key feature of the family of LML detectors is to search out an LML-J vector with neighborhood size J. Hence, while there are various variations, many of the abovementioned local search detectors belong to the family of LAS detectors and/or the family of LML detectors.

It is not surprising that the LAS detector and other local search detectors in the literature are successful, and approach the GML performance in the massive antenna MIMO channels [26–58]. The LRS CDMA channel [21, 22], the sparse LRS CDMA channel [23, 24], and the massive antenna MIMO channel [27] are particular instances of the general large MIMO channel [25]. These large MIMO channels have a common property: as the dimension of channel vectors tends to infinity, the cross correlation coefficient between two channel vectors converges to zero. This phenomenon is termed "channel hardening" for massive antenna MIMO channels [27]. It is because of this property that the general large MIMO channel possesses the "LML characteristic." Therefore, as prophesied in Refs. [21, 22], in a practical range of channel load and SNR, "there is typically no LML but the GML point. The likelihood function is sufficiently smooth. Thus, the linear-complex WSLAS detector (and other detectors) with local search can reach the GML point that is usually NP-hard to obtain. All the results are also applicable to the LML detectors with neighborhood sizes greater than one [15–17]."

As the number of mobile devices and the required spectral efficiency ever increase [59, 60], the next-generation mobile network will rely on massive MIMO communications to fulfill the ever-increasing demand, and then the LAS and LML detectors are expected to play a role. To this end, it is essential to understand the LML and LAS detectors and their properties. In this book chapter, we review the family of LML detectors and the family of LAS detectors in a unified framework. The focus is on their formulation, relationships, properties, and the GML BER and spectral efficiency in massive MIMO channels.

The rest of the book chapter is organized as follows. Section 2 presents the GML detector and the family of LML detectors. Section 3 addresses the family of LAS detectors and their performance in BER and AME. Section 4 studies the LML characteristic of massive MIMO channels and the analytical BER of the WSLAS and LML detectors by the replica method. Section 5 briefly discusses the application of LML and LAS families to the next generation of multiaccess 5G/6G and beyond wireless networks. Conclusions are drawn in Section 6.

2. The family of local maximum likelihood (LML) detectors

2.1 The MIMO channel model

Consider a general MIMO channel where K complex symbols are transmitted through a complex Gaussian channel. Each vector of K symbols is equiprobably selected from an alphabet Ω_K. The kth symbol b_k is modulated by an N-dimensional complex channel vector $\boldsymbol{s}_k$ with unit length $\|\boldsymbol{s}_k\| = 1$. The energy of the kth symbol is A_k^2. The receiver receives an N-dimensional complex vector

$$\boldsymbol{r} = \mathbf{S}\mathbf{A}\boldsymbol{b} + \boldsymbol{m} \tag{1}$$

where $\mathbf{S} = (\mathbf{s}_1, \dots, \mathbf{s}_K) \in C^{N\times K}$ is the channel matrix, $\mathbf{A} = \mathrm{diag}(A_1, \dots, A_K)$ is the diagonal matrix of symbol amplitudes, $\boldsymbol{b} = (b_1, \dots, b_K)^T$ is the vector of symbols, and $\boldsymbol{m} \sim CN(\mathbf{0}, \sigma^2\mathbf{I}_N)$ is the AWGN.

A K-dimensional sufficient statistic can be obtained by the matched filter (MF) $\mathbf{S}^H$ as

$$\boldsymbol{y} = \mathbf{S}^H\boldsymbol{r} = \mathbf{R}\mathbf{A}\boldsymbol{b} + \boldsymbol{n} \tag{2}$$

where $\mathbf{R} = \mathbf{S}^H\mathbf{S}$ is the crosscorrelation matrix of channel vectors and $\boldsymbol{n} = \mathbf{S}^H\boldsymbol{m} \sim CN(\mathbf{0}, \sigma^2\mathbf{R})$ with H denoting the Hermitian transpose is the noise at the MF output. The probability density function of $\boldsymbol{y}$ given $\boldsymbol{b}$, $\mathbf{S}$, $\mathbf{A}$ is

$$p(\boldsymbol{y}|\boldsymbol{b}, \mathbf{S}) = \frac{1}{(\pi\sigma^2|\mathbf{R}|)^K} \exp\left(-\frac{(\boldsymbol{y} - \mathbf{R}\mathbf{A}\boldsymbol{b})^H\mathbf{R}^{-1}(\boldsymbol{y} - \mathbf{R}\mathbf{A}\boldsymbol{b})}{\sigma^2}\right) \tag{3}$$

where $\|\bullet\|$ is the l_2 norm. When $\mathbf{R}$ is not invertible, $\mathbf{R}^{-1}$ denotes its pseudoinverse. The posterior probability of $\boldsymbol{b}$ is

$$p(\boldsymbol{b}|\boldsymbol{y}, \mathbf{S}) = \frac{p(\boldsymbol{b})p(\boldsymbol{y}|\boldsymbol{b}, \mathbf{S})}{\sum_{\boldsymbol{a}\in\Omega_K} p(\boldsymbol{a})p(\boldsymbol{y}|\boldsymbol{a}, \mathbf{S})}. \tag{4}$$

We will use both $\boldsymbol{r}$ and $\boldsymbol{y}$ henceforth whenever one is more suitable.

The general MIMO channel model in Eq. (1) includes several typical MIMO communication systems. First, multiple users, each equipped with one or multiple transmit antennas, access a base station (BS) that is equipped with N receive antennas [27]. Second, in a CDMA system where K users access a BS with one antenna, $\boldsymbol{s}_k$ represents the spreading sequence of the kth user and N is the spreading factor [18, 19]. Third, by the multicode technique with extended symbol periods, each user can transmit multiple symbols with/without spread spectrum and/or time for single or multiple users [21, 22]. The codes can be long, short, and sparse [22–24]. Fourth, the multi-transmit antenna, multi-receive antenna, CDMA, and multicode techniques can be applied in combination, and each yields the same channel model in Eq. (1).

The GML detector, the family of LML detectors, and the family of LAS detectors can all be applied to any complex symbol constellation. As often considered in the literature, we shall subsequently focus on the BPSK signals transmitted through a real AWGN channel. Each symbol independently and equiprobably takes on $\{-1, 1\}$ and then $\boldsymbol{b} \in \Omega_K \equiv \{-1, 1\}^K$. The channel noise is $\boldsymbol{m} \sim N(\mathbf{0}, \sigma^2 \mathbf{I}_N)$. Accordingly, the probability density function of $\boldsymbol{r}$ conditioned on $(\boldsymbol{b}, \mathbf{S})$ is obtained in Eq. (3) by replacing σ^2 and N with $2\sigma^2$ and $N/2$, respectively. $\mathbf{S}$, $\boldsymbol{b}$, and $\boldsymbol{m}$ are assumed to be mutually independent. The prior probability distribution of $\boldsymbol{b}$ is $p(\boldsymbol{b}) = 2^{-K}$, $\forall \boldsymbol{b} \in \Omega_K$. The average symbol energy is normalized to unit $K^{-1}\sum_{k=1}^{K} A_k^2 = 1$. The channel load is $\alpha = K/N$ in the unit of bits/dimension. In a system where the channel dimension is the product of time and bandwidth without spatial diversity, the channel load is in the unit of bits/s/Hz. Such a system can be the CDMA system where the channel vectors are code sequences with/without spread spectrum and/or time [21, 22].

2.2 The global maximum likelihood (GML) detector

A bit vector $\boldsymbol{b}$ can be demodulated from the received signal $\boldsymbol{r}$. With equiprobable bits, the probability density function $p(\boldsymbol{y}|\boldsymbol{b}, \mathbf{S})$ is also a likelihood function of $\boldsymbol{b}$ with given the signal $\boldsymbol{y}$. Equivalently, a likelihood function can be defined from Eq. (3) as

$$f(\boldsymbol{y}|\boldsymbol{b}) = -\frac{1}{2}(\boldsymbol{y} - \mathbf{R}\mathbf{A}\boldsymbol{b})^H \mathbf{R}^{-1}(\boldsymbol{y} - \mathbf{R}\mathbf{A}\boldsymbol{b}). \tag{5}$$

The GML detector φ^{GML} selects the bit vector that achieves the maximum likelihood among all possible vectors

$$\hat{\boldsymbol{b}} = \arg\max_{\boldsymbol{b} \in \Omega_K} f(\boldsymbol{y}|\boldsymbol{b}), \forall \boldsymbol{y}. \tag{6}$$

With equiprobable bits, the GML detector also achieves the global maximum posterior probability $\hat{\boldsymbol{b}} = \arg\max_{\boldsymbol{b} \in \Omega_K} p(\boldsymbol{b}|\boldsymbol{y}, \mathbf{S})$.

By maximizing the likelihood function globally over Ω_K for any $\boldsymbol{r}$, the GML detector minimizes the probability of error $P_e(\varphi) \equiv \Pr(\boldsymbol{b}^{\varphi} \neq \boldsymbol{b})$ among all detectors φ, and so is the optimum detector. On the other hand, the GML detector needs to compare the likelihood of all 2^K vectors in Ω_K and, therefore, is NP-hard in computation. The GML detector is usually infeasible in practical communication systems when K is large.

2.3 Local maximum likelihood (LML) detectors

To trade off error performance for a low computational complexity, the LML detectors can be applied [15–17]. An LML detector depends on the neighborhood size of a bit vector. The neighborhood of $\boldsymbol{b}$ with neighborhood size J is defined by the set of vectors that differ from $\boldsymbol{b}$ by at most J bits

$$\Omega_J(\boldsymbol{b}) = \{\boldsymbol{a} \in \Omega_K | \|\boldsymbol{a} - \boldsymbol{b}\|_1/2 \leq J\}$$

where $\|\bullet\|_1$ is the l_1 norm.

Given $\boldsymbol{y}$, an LML detector φ_J^{LML} with neighborhood size J chooses a vector $\hat{\boldsymbol{b}}$ that attains the maximum likelihood in its neighborhood of size J [15–17]

$$\hat{\boldsymbol{b}} = \arg \max_{\boldsymbol{b} \in \Omega_J(\hat{\boldsymbol{b}})} f(\boldsymbol{y}|\boldsymbol{b}), \forall \boldsymbol{y}. \tag{7}$$

With the equiprobable bits, the LML detector also achieves the local maximum posterior probability $\hat{\boldsymbol{b}} = \arg\max_{\boldsymbol{b} \in \Omega_J(\hat{\boldsymbol{b}})} p(\boldsymbol{b}|\boldsymbol{r}, \mathbf{S})$.

In the particular case of $J = K$, the neighborhood of a vector is the entire set of all vectors $\Omega_J\left(\hat{\boldsymbol{b}}\right) = \Omega_K$. The LML detector in Eq. (7) becomes the GML detector. Henceforth, for simplicity, LML-J denotes LML with neighborhood size J.

An LML-J vector for $J < K$ is a fixed point of Eq. (7), and its righthand side has a neighborhood depending on the vector $\hat{\boldsymbol{b}}$ to be determined. In contrast, the GML (i.e., LML-K) vector in Eq. (6) has the neighborhood of the entire vector set independent of the vector to be determined.

2.4 GML and LML regions

The performance of the LML and GML detectors can be understood by the LML and GML vectors and their decision regions.

Given $\boldsymbol{y}$, an LML-J vector solves Eq. (7) and the set of LML vectors with neighborhood size J is defined as [15–17].

$$\Psi_J^{\text{LML}}(\boldsymbol{y}) = \{\boldsymbol{b} \in \Omega_K | f(\boldsymbol{y}|\boldsymbol{b}) \geq f(\boldsymbol{y}|\boldsymbol{a}), \forall \boldsymbol{a} \in \Omega_J \backslash \{\boldsymbol{b}\}\}. \tag{8}$$

In particular, the set of GML vectors is

$$\Psi^{\text{GML}}(\boldsymbol{y}) = \Psi_K^{\text{LML}}(\boldsymbol{y}) = \{\boldsymbol{b} \in \Omega_K | f(\boldsymbol{y}|\boldsymbol{b}) \geq f(\boldsymbol{y}|\boldsymbol{a}), \forall \boldsymbol{a} \in \Omega_K \backslash \{\boldsymbol{b}\}\}.$$

Depending on $\boldsymbol{y}$, there might be multiple vectors in $\Psi_J^{\text{LML}}(\boldsymbol{y})$ that solve Eq. (7). On the other hand, a unique GML vector solves Eq. (6), that is, $\Psi^{\text{GML}}(\boldsymbol{y}) = \left\{\boldsymbol{b}^{\text{GML}}(\boldsymbol{y})\right\}$ unless $\boldsymbol{y}$ is on the boundary of multiple decision regions, which occurs with probability zero. Hence, the GML vector is unique in general.

Since an LML vector with a neighborhood size $J + 1$ is also an LML vector with a neighborhood size J, the following relationships hold

$$\Psi^{\text{GML}}(\boldsymbol{y}) \subseteq \Psi_{K-1}^{\text{LML}}(\boldsymbol{y}) \subseteq \cdots \subseteq \Psi_2^{\text{LML}}(\boldsymbol{y}) \subseteq \Psi_1^{\text{LML}}(\boldsymbol{y}). \tag{9}$$

Hence, an LML-J vector is also an LML-M vector for $M < J$. The GML vector is an LML-J vector for all $J < K$.

The relationships in Eq. (9) imply two properties of the family of LML detectors. First, it is evident that all the equalities in Eq. (9) hold with probability one (i.e., $\forall \boldsymbol{y}$ except the boundary of decision regions) iff in the trivial orthogonal channel where the channel vectors $\boldsymbol{s}_k$ are orthogonal. When there are multiple LML-J vectors in $\Psi_J^{\text{LML}}(\boldsymbol{y})$, an LML-$J$ detector can output any LML-J vector and, therefore, the LML detector φ_J^{LML} is not unique for $J < K$. Second, when there are multiple LML-J vectors, though all the LML-J vectors in $\Psi_J^{\text{LML}}(\boldsymbol{y})$ solve Eq. (7), some of the LML-J vectors have a higher likelihood than the others. For $\boldsymbol{a}, \boldsymbol{b} \in \Psi_J^{\text{LML}}(\boldsymbol{y})$, if $\boldsymbol{a} \in \Psi_{J+1}^{\text{LML}}(\boldsymbol{y})$ but $\boldsymbol{b} \notin \Psi_{J+1}^{\text{LML}}(\boldsymbol{y})$, then the likelihood of $\boldsymbol{a}$ is higher than $\boldsymbol{b}$. In other words, in addition to the LML-$(J+1)$ vectors in $\Psi_{J+1}^{\text{LML}}(\boldsymbol{y})$, $\Psi_J^{\text{LML}}(\boldsymbol{y})$ contains some additional LML-J vectors whose likelihoods are lower than the likelihoods of all LML-$(J+1)$ vectors in $\Psi_{J+1}^{\text{LML}}(\boldsymbol{y})$. An LML-$J$ detector φ_J^{LML} selects one vector in $\Psi_J^{\text{LML}}(\boldsymbol{y})$ and achieves a probability of error not less than an LML-$(J+1)$ detector. Hence, the following relationships hold

$$P_e\left(\varphi^{\text{GML}}\right) \le P_e\left(\varphi_{K-1}^{\text{LML}}\right) \le \dots \le P_e\left(\varphi_2^{\text{LML}}\right) \le P_e\left(\varphi_1^{\text{LML}}\right). \tag{10}$$

Note that in Eq. (10), the equality in $P_e\left(\varphi_{J+1}^{\text{LML}}\right) \le P_e\left(\varphi_J^{\text{LML}}\right)$ holds iff φ_J^{LML} chooses only the vectors belonging to $\Psi_{J+1}^{\text{LML}}(\boldsymbol{y})$ among all vectors in $\Psi_J^{\text{LML}}(\boldsymbol{y})$; but this means that φ_J^{LML} is an LML-$(J+1)$ detector. Hence, all the inequalities in Eq. (10) hold in general, except for the trial orthogonal channel.

The LML-J region of a bit vector $\boldsymbol{b} \in \Omega_K$ is defined as the region of $\boldsymbol{y} \in \mathbb{R}^K$ such that $\boldsymbol{b}$ is an LML-J vector [15–17].

$$V_J^{\text{LML}}(\boldsymbol{b}) = \left\{\boldsymbol{y} \in \mathbb{R}^K | f(\boldsymbol{y}|\boldsymbol{b}) \ge f(\boldsymbol{y}|\boldsymbol{a}), \forall \boldsymbol{a} \in \Omega_J \backslash \{\boldsymbol{b}\}\right\}. \tag{11}$$

In particular, the GML region is the LML-K region

$$V^{\text{GML}}(\boldsymbol{b}) = V_K^{\text{LML}}(\boldsymbol{b}) = \left\{\boldsymbol{y} \in \mathbb{R}^K | f(\boldsymbol{y}|\boldsymbol{b}) \ge f(\boldsymbol{y}|\boldsymbol{a}), \forall \boldsymbol{a} \in \Omega_K \backslash \{\boldsymbol{b}\}\right\}.$$

For any $\boldsymbol{y} \in V_{J+1}^{\text{LML}}(\boldsymbol{b})$, $\boldsymbol{b}$ is an LML-$(J+1)$ vector. Then $\boldsymbol{b}$ is also an LML-J vector. Consequently, $\boldsymbol{y} \in V_J^{\text{LML}}(\boldsymbol{b})$. Hence, the following relationships hold

$$V^{\text{GML}}(\boldsymbol{b}) \subseteq V_{K-1}^{\text{LML}}(\boldsymbol{b}) \subseteq \cdots \subseteq V_2^{\text{LML}}(\boldsymbol{b}) \subseteq V_1^{\text{LML}}(\boldsymbol{b}). \tag{12}$$

In the region of $V_J^{\text{LML}}(\boldsymbol{b}) \backslash V_{J+1}^{\text{LML}}(\boldsymbol{b})$, $\boldsymbol{b}$ is an LML-J vector but is not an LML-$(J+1)$ vector. In other words, in addition to $\text{V}_{J+1}^{\text{LML}}(\boldsymbol{b})$, $\text{V}_J^{\text{LML}}(\boldsymbol{b})$ contains an extra region where $\boldsymbol{b}$ is an LML-J vector and $\boldsymbol{b}$ has a likelihood not greater than the other LML-$(J+1)$ vectors. This also implies that Eq. (10) is true. Hence, as the neighborhood size J increases, the error probability of an LML-J detector monotonically decreases.

Figure 1(a) shows the LML and GML regions in **Figure 1** with neighborhood size $J = 1$ and $J = 2$, respectively, in a two-bit channel with $R_{12} = \rho > 0$. A more detailed explanation of the LML and GML regions in **Figure 1** is presented beneath Eq. (32) in Section 3.5.

Given $K \ge 3$, as the neighborhood size J increases, the region where multiple LML-J vectors coexist shrinks, and then the BER performance of an LML-J detector improves. In general, the GML detector performs substantially better than an LML-J

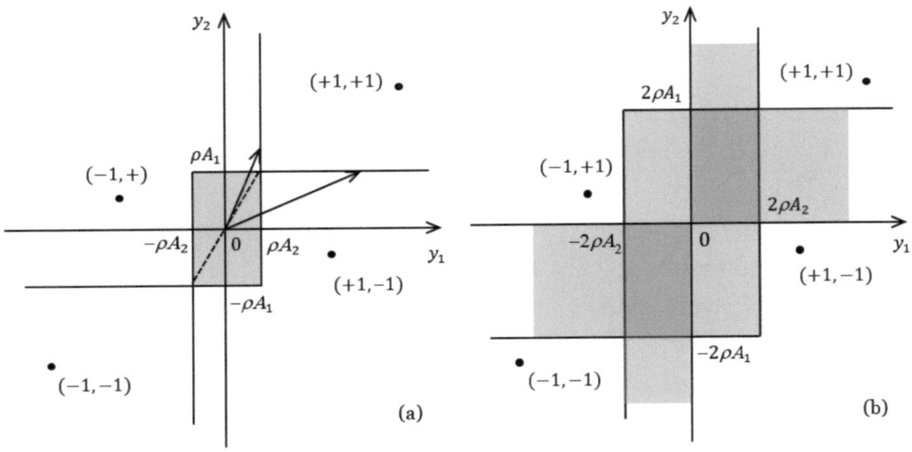

Figure 1.
*The **y** space for the two-bit channel with $\rho = 0.4$, $A_1 = 1$ and $A_2 = 0.6$. (a) The fixed-point regions (or LML-1 regions) of the WSLAS detectors, and (b) The fixed-point regions of the PLAS detector.*

detector with a small J in a small- or median-size system [15–17]. However, as presented in Section 4, in a large MIMO system where both K and N tend to infinity at the same rate, an LML-J detector with a small J can perform equally well as the GML detector in a wide range of channel load and SNR [22, 25].

2.5 The family of LMLAS detector

The family of local-maximum-likelihood likelihood-ascent-search (LMLAS) detectors [15–17] can achieve the LML-J detection for $J \in \{1, \ldots, K\}$. For simplicity, LMLAS-J means LMLAS with neighborhood size J.

The family of LMLAS detectors is designed in principle upon Eq. (7). Since both sides of Eq. (7) depend on $\hat{\boldsymbol{b}}$, the LML-J vector $\hat{\boldsymbol{b}}$ is a fixed point that solves Eq. (7). The definition of the LML-J detector in Eq. (7) suggests that an LML-J vector can be obtained by an iterative search [15–17]. Given $\boldsymbol{y}$, starting with an initial vector $\boldsymbol{b}(0)$, search for a higher-likelihood vector $\boldsymbol{b}(1)$ in $\Omega_J(\boldsymbol{b}(0))$, the neighborhood of $\boldsymbol{b}(0)$ with neighborhood size J; if $\boldsymbol{b}(1)$ does not attain the maximum likelihood in its neighborhood $\Omega_J(\boldsymbol{b}(1))$, then search for a higher-likelihood vector $\boldsymbol{b}(2) \in \Omega_J(\boldsymbol{b}(1))$; repeat the search until an LML-J vector $\hat{\boldsymbol{b}}$ is obtained, which achieves the maximum likelihood in its own neighborhood $\Omega_J\left(\hat{\boldsymbol{b}}\right)$. To remove redundant computations, the search is guided by the likelihood gradient of the current vector, and the gradient is updated computationally efficiently by Eq. (18) in the next section.

2.6 Computational complexity

An LMLAS detector can computationally efficiently perform the likelihood ascent search [15–17]. To avoid redundant computation, the LMLAS detectors employ the gradient of the likelihood function $f(\boldsymbol{y}|\boldsymbol{b})$. The likelihood of a vector can be compared with the likelihood of the current vector in terms of the gradient of likelihood function at the current vector. Once a vector with a higher likelihood is accepted as the current vector, the gradient can be updated by parallel vector additions of Eq. (18). Since the likelihood comparison and the gradient update are two core computations and both can be computationally efficiently performed, the family of LMLAS detectors achieves high efficiency in computation. To determine an LML-J vector, it needs to compare the likelihood of the current vector with other vectors in the neighborhood of size J. Thus, the per-bit computational complexity of an LMLAS-J detector is in the order of Refs. [15–17]

$$C_J^{\mathrm{LML}} = O\left(\sum_{j=1}^{J} \binom{K}{j}\right), \tag{13}$$

and, therefore,

$$C^{GML} = C_K^{\mathrm{LML}} > \ldots > C_2^{\mathrm{LML}} > C_1^{\mathrm{LML}}. \tag{14}$$

In particular, the LMLAS-1 detector achieves the lowest computational complexity $C_1^{\mathrm{LML}} = O(K)$ linear in K. The LMLAS-K detector (i.e., the GML detector) is the most complex with a computational complexity $C^{GML} = C_K^{\mathrm{LML}} = O(2^K)$ exponential in K.

3. The family of likelihood ascent search (LAS) detectors

By choosing a different neighborhood size J, the family of LMLAS detectors can achieve a different tradeoff of computational complexity and error performance. An LMLAS-1 detector achieves the lowest computational complexity and the worst error performance. To enable further reduction of computational complexity, the family of LAS detectors can be applied [11–14, 18, 20–24].

3.1 Criteria for the updating rule

The idea of designing the family of LAS detectors is to search a sequence of bit vectors $\boldsymbol{b}(n)$ such that the likelihood of $\boldsymbol{b}(n)$ monotonically increases with the step n until reaching a fixed point. The bits scheduled to update in a step are called update candidates. Updating a bit is to check the flip condition of the bit, but it may or may not result in a bit flip according to the updating rule.

Suppose $L(n) \subseteq \{1, \ldots, K\}$ is the index set of update candidates in the nth step. The gradient $\nabla f(\boldsymbol{b})$ of the likelihood function with respect to $\boldsymbol{b}$ evaluated at $\boldsymbol{b}(n)$ equals

$$\boldsymbol{g}(n) = -\mathbf{H}\boldsymbol{b}(n) + \mathbf{A}\boldsymbol{y} \tag{15}$$

with $\mathbf{H} = \mathbf{ARA}$ whose diagonal elements are the symbol energies $H_{kk} = A_k^2$ as $R_{kk} = 1$ for any k. $\boldsymbol{g}(n)$ is simply equal to $\mathbf{A}$ times the difference between the MF output and the signal reconstructed by $\boldsymbol{b}(n)$, i.e., $\boldsymbol{g}(n) = \mathbf{A}[-\mathbf{RA}\boldsymbol{b}(n) + \boldsymbol{y}]$. The kth component of the likelihood gradient is

$$g_k(n) = -\sum_{i=1}^{K} H_{ki} b_i(n) + A_k y_k. \tag{16}$$

If the bits in $L_p(n) \subseteq L(n)$ are flipped in the step n, then the new bit vector is

$$\boldsymbol{b}(n+1) = \boldsymbol{b}(n) - 2\sum_{k \in L_p(n)} b_k(n)\boldsymbol{e}_k \tag{17}$$

where the kth element of $\boldsymbol{e}_k$ is equal to one, and others are zero. The likelihood gradient in the next step can be efficiently updated by

$$\boldsymbol{g}(n+1) = \boldsymbol{g}(n) + 2\sum_{k \in L_p(n)} b_k(n)\boldsymbol{h}_k \tag{18}$$

where $\boldsymbol{h}_k$ is the kth column vector of $\mathbf{H}$. The likelihood change $\Delta f(n) = f[\boldsymbol{b}(n+1)] - f[\boldsymbol{b}(n)]$ can be calculated in terms of $\boldsymbol{g}(n)$ as

$$\Delta f(n) = \Delta\boldsymbol{b}^T(n)\left[\boldsymbol{g}(n) + \frac{1}{2}\boldsymbol{z}(n)\right] \tag{19}$$

with $\Delta\boldsymbol{b}(n) = \boldsymbol{b}(n+1) - \boldsymbol{b}(n)$ and $\boldsymbol{z}(n) = -\mathbf{H}\Delta\boldsymbol{b}(n)$.

The LAS detector is developed based on the criteria that (i) the updating rule is computationally efficient, (ii) the new bit vector $\boldsymbol{b}(n+1)$ must have a higher likelihood than $\boldsymbol{b}(n)$ if $\boldsymbol{b}(n+1) \neq \boldsymbol{b}(n)$, and (iii) under the same framework of an updating rule, $\boldsymbol{b}(n+1)$ has the highest likelihood. It is shown that a LAS detector satisfies these criteria [11–14, 18].

3.2 The LAS detectors

The following generalized LAS detector applies to all possible sequences of candidate sets $L(n)$, $n \geq 0$ and defines the family of LAS detectors.

LAS detector: Given $L(n) \subseteq \{1, \dots, K\}$ for all $n \geq 0$ and an initial vector $\boldsymbol{b}(0) \in \Omega_K$. At step n all the bits for $k \in L(n)$ are updated by

$$b_k(n+1) = \begin{cases} +1, & \text{if } b_k(n) = -1 \text{ and } g_k(n) > t_k(n), \\ -1, & \text{if } b_k(n) = +1 \text{ and } g_k(n) < -t_k(n), \\ b_k(n), & \text{otherwise,} \end{cases} \tag{20}$$

where the kth threshold is determined by the elements of $\mathbf{H}$ as

$$t_k(n) = \sum_{j \in L(n)} |h_{kj}|, \tag{21}$$

all the bits for $k \notin L(n)$ remain unchanged $b_k(n+1) = b_k(n)$, and then $\boldsymbol{g}(n+1)$ is updated by Eq. (18) in which $L_p(n)$ is the index set of flipped bits in Eq. (20). $\boldsymbol{b}^*$ is the final demodulated vector if $\boldsymbol{b}(n) = \boldsymbol{b}^*$, $\forall n \geq n^*$ with some $n^* \geq 0$.

In the LAS update rule of Eq. (20), if the thresholds are too low, the iteration will enter a limit cycle. On the other hand, if the thresholds are too high, after updating, the likelihood of $\boldsymbol{b}(n+1)$ is not sufficiently high, and the search will stop at a bit vector with a low likelihood. The following theorem indicates that the thresholds in the LAS detector are optimum [12–14].

Theorem 1: For any $L(n) \subseteq \{1, \dots, K\}$ and $\mathbf{H}$, the thresholds $t_k(n)$ in Eq. (21) are necessary and sufficient for the LAS detector to increase the likelihood of a nonzero update $\boldsymbol{b}(n+1) \neq \boldsymbol{b}(n)$.

In practice, $L(n)$ for $n \geq 0$ can be scheduled so that all the bits are finally periodically updated. If no bit is flipped in a period, the LAS detector has reached a fixed point $\boldsymbol{b}^*$ and shall terminate. It is assumed that every bit is updated once more without a flip before reaching a fixed point. The sequence of $L(n)$ is deterministic in most cases, though it can be random. The LAS detector is step-invariant if $L(n_1) = L(n_2)$ for all $\boldsymbol{b}(n_1) = \boldsymbol{b}(n_2)$, $n_1 \neq n_2$.

Specifying a sequence of $L(n)$for $n \geq 0$, one determines a particular LAS detector. One of the most straightforward sequences is to update one bit in each step, which produces a sequential LAS (SLAS) detector with the lowest and step-invariant thresholds

$$t_k = A_k^2. \tag{22}$$

The SLAS detector can update the bits in a circular or random order.

All the SLAS detectors belong to the larger set of wide-sense SLAS (WSLAS) detectors that set $|L(n)| = 1$, $\forall n \geq n'$ with some $n' \geq 0$. The WSLAS detector eventually updates one bit in each step and works in the SLAS mode. All the WSLAS detectors, including the SLAS detector, converge to an LML-1 vector and, therefore, are the LML-1 detectors [11–14].

Another simplest sequence is $L(n) = \{1, \ldots, K\}$, $\forall n \geq 0$, which yields the parallel LAS (PLAS) detector with the highest and step-invariant thresholds

$$t_k = \sum_{j=1}^{K} |h_{kj}|. \tag{23}$$

The PLAS detector updates all K bits in each step.

The PLAS and the SLAS detectors are particular instances of the group-parallel LAS (GPLAS) detectors that update bits group by group. If ζ is a collection of subsets that partitions $\{1, \ldots K\}$, a GPLAS detector has $L(n) \in \zeta$ with thresholds

$$t_k = \sum_{j \in L} |h_{kj}|, \forall k \in L \in \zeta. \tag{24}$$

More complicated sequences of $L(n)$, $n \geq 0$ can be specified, such as those WSLAS detectors based on the EHE and FMD criteria [11] to converge to a fixed point with a higher likelihood in fewer steps.

An initial vector $\boldsymbol{b}(0)$ with a lower error probability can make the LAS detector converge faster to a fixed point of lower error probability. To reduce dependency on the initial computational cost, a random vector in Ω_K and the MF detector output $\boldsymbol{b}(0) = \mathrm{sgn}(\boldsymbol{y})$ can be employed as the initial vector.

The updating condition in Eq. (20) can be concisely written as $b_k(n)g_k(n) < -t_k(n)$ but it is elaborately written in the form of Eq. (20) for convenience of hardware implementation. Though Eq. (20) can also be rewritten without the use of the likelihood gradient $\boldsymbol{g}(n)$, there are good reasons to use it. First, the efficient updating of $\boldsymbol{g}(n)$ in Eq. (18) is the key to reducing redundant computations. Second, Eq. (18) can be implemented on hardware suitable for fast parallel computation like a multilayer perceptron. Third, searching the next vector along $\boldsymbol{g}(n)$ with likelihood ascent is the motivation of LAS design, which identifies the LAS detectors from the other detectors motivated by different criteria [61–64].

3.3 Monotonic likelihood ascent and stability

An iterative algorithm that searches out a sequence of bit vectors is stable if it monotonically increases an upper-bounded Lyapunov function, thus ensuring the convergence to a fixed point in a finite number of iterations. Otherwise, it enters a limit cycle on which the Lyapunov function is not monotonic. The likelihood function $f(\boldsymbol{y}|\boldsymbol{b})$ in Eq. (5) is a Lyapunov function of the LAS detector. The following theorem is proved [11–14, 18].

Theorem 2: Consider a LAS detector with $L(n)$, $n \geq 0$ that generates a sequence of bit vectors $\boldsymbol{b}(n)$, $n \geq 0$. (i) For any $\boldsymbol{y}, f[\boldsymbol{y}|\boldsymbol{b}(n+1)] \geq f[\boldsymbol{y}|\boldsymbol{b}(n)]$, $\forall n \geq 0$ with equality iff $\boldsymbol{b}(n+1) = \boldsymbol{b}(n)$; (ii) $\boldsymbol{b}(n)$ converges to a fixed point $\boldsymbol{b}^*$, that is, $\boldsymbol{b}(n) = \boldsymbol{b}^*$, $\forall n \geq n^*$ with a finite number of steps $n^* \geq 0$; (iii) $P_e[\boldsymbol{b}(n+1)] \leq P_e[\boldsymbol{b}(n)]$ with equality iff $\Pr[\boldsymbol{b}(n+1) = \boldsymbol{b}(n)] = 1$; (iv) $P_e(\boldsymbol{b}^*) \leq P_e[\boldsymbol{b}(0)]$ with equality iff $\boldsymbol{b}(0)$ is a fixed point of the LAS detector with probability one, and for the WSLAS detector with equality iff $\boldsymbol{b}(0)$ is an LML-1 vector with probability one.

The theorem indicates that a LAS detector monotonically increases the likelihood, is stable, and monotonically reduces the probability of error for the sequence of searched vectors. Unless the initial detector is a fixed point with probability one, the LAS detector reduces the error probability of the initial detector.

3.4 Tradeoff between performance and computational complexity

The LAS detector can achieve a tradeoff between the error performance and the computational complexity by a proper design of the sequence of $L(n)$ for $n \geq 0$, the bits to be updated in each step. The following theorem is obtained [12]. The first inequality is obtained from the updating rule in Eq. (20) and the second inequality is further obtained from the fact of $f(\boldsymbol{y}|\boldsymbol{b}) = -\frac{1}{2}\boldsymbol{g}^H\mathbf{H}^{-1}\boldsymbol{g}$.

Theorem 3: Suppose a LAS detector converges to a fixed point $\boldsymbol{b}^*$ after n^* steps, $\bigcup_{n \geq n^*} L(n) = \{1, \ldots, K\}$, and $\boldsymbol{b}^*$ is not bounded. At $\boldsymbol{b}^*$ the gradient of the likelihood function is upper bound by

$$\|\boldsymbol{g}^*\|_1 \leq \sum_{k=1}^{K} \min_{n \geq n^*, k \in L(n)} \{t_k(n)\} \tag{25}$$

and the likelihood is lower bounded by

$$f(\boldsymbol{y}|\boldsymbol{b}^*) \geq -\frac{1}{2}\|\mathbf{H}^{-1}\| \sum_{k=1}^{K} \min_{n \geq n^*, k \in L(n)} \{t_k^2(n)\}. \tag{26}$$

Theorem 3 implies that the fewer bits that are updated at each step after reaching a fixed point (i.e., $|L(n)|$ for $n \geq n^*$ is smaller), the lower the thresholds, and then the lower the upper bounds of $\|\boldsymbol{g}^*\|_1$ and the higher the lower bound on the likelihood $f(\boldsymbol{y}|\boldsymbol{b}^*)$. In other words, a smaller $|L(n)|$ for $n \geq n^*$ results in a higher likelihood of the demodulated bit vector $\boldsymbol{b}^*$.

On the other hand, the more bits to be updated at a step, then the larger the search region in Ω_K (i.e., a possible larger $\|\boldsymbol{b}(n+1) - \boldsymbol{b}(n)\|_1$), the higher the thresholds, the fewer steps to converge to a fixed point. A convergence in fewer steps is particularly preferred when the LAS detector is implemented on hardware like a multilayer perceptron suitable for parallel computations where the computational time for demodulating a bit vector depends on the number of steps instead of the total number of additions and multiplications.

Hence, to trade off the error performance and the computational time, a better design of $L(n)$ for $n \geq 0$ is to update more bits with a large $|L(n)|$ in a few initial steps to achieve fast convergence and then gradually decrease the number of updated bits with a smaller $|L(n)|$, say $|L(n)| = 1$, to achieve the highest likelihood of $\boldsymbol{b}^*$ – an LML-1 vector, which is the criterion in the design of $L(n)$, $n \geq 0$ for a WSLAS detector [11–15, 18].

The updating rule Eq. (20) with the gradient update Eq. (18) is the core cost of computational complexity and is the key to the linear complexity of the LAS detector. The total number of bit flips from $\boldsymbol{b}(0)$ to $\boldsymbol{b}^*$ equals $M = \sum_{n=0}^{n^*} |L_p(n)|$, which depends on $L(n)$, $\boldsymbol{b}(0)$, and $\boldsymbol{y}$, and is random. Define the bit flip rate (BFR) as the average number of flips per bit $c = E(M)/K$. Computational complexity can be defined as the average number of additions per bit. Each bit flip results in K additions in Eq. (18). The computational complexity equals $E(KM)/K = E(M) = cK$ linear in K. The BFR c is about 0.5 as analyzed in [12–14], and less than 0.5 as observed in all simulations in various conditions, part of which is reported in [11, 21–24]. With the implementation of hardware suitable for vector additions, the computational complexity is further reduced by a factor of K.

3.5 Fixed-point region

The error performance of the LAS detector is determined by its output fixed point. Given $L(n)$ for $n \geq 0$, in the $\mathbf{y}$ space, there is a region of $\boldsymbol{b}$ associated with the initial detector $\boldsymbol{b}(0)$ where the LAS detector converges from $\boldsymbol{b}(0)$ to $\boldsymbol{b}$. To determine the effect of a particular sequence $L(n)$ on error performance, define the fixed-point region of $\boldsymbol{b}$ by

$$V^{\text{LAS}}(\boldsymbol{b}) \equiv \{\boldsymbol{y} \in \mathbb{R}^K | \text{Exist } \boldsymbol{b}(0) \text{ s.t. the LAS detector converges to } \boldsymbol{b}\},$$

which is the union of such regions associated with all initial detectors. Conversely, given $\boldsymbol{y}$, define the set of fixed points

$$\Psi^{\text{LAS}}(\boldsymbol{y}) \equiv \{\boldsymbol{b} \in \Omega_K | \text{Exist } \boldsymbol{b}(0), L(n), n \geq 0 \text{ s.t. the LAS detector converges to } \boldsymbol{b}\}.$$

The following proposition is obtained [18].

Proposition 1: For any $\boldsymbol{y}$, denote by $\Lambda(\boldsymbol{y}, \boldsymbol{b}) = \{\boldsymbol{b}(0) \in \Omega_K | \text{LAS} : \boldsymbol{y}, \boldsymbol{b}(0) \to \boldsymbol{b}\}$ the set of initial bit vectors from which the LAS detector converges to $\mathbf{b}$. Let $t_k^* = \max_{\boldsymbol{b}(0) \in \Lambda(\boldsymbol{y}, \boldsymbol{b})} \min_{n \geq n_k^*[\boldsymbol{b}(0)], k \in L(n)} t_k(n)$ where $n_k^*[\boldsymbol{b}(0)]$ is the last flip step of the kth bit with the initial $\boldsymbol{b}(0)$. Then

$$V^{\text{LAS}}(\boldsymbol{b}) = \{\boldsymbol{y} \in \mathbb{R}^K | \boldsymbol{b} \otimes (\mathbf{A}\boldsymbol{y} - \mathbf{H}\boldsymbol{b}) \geq -\boldsymbol{t}^*\} \tag{27}$$

and conversely

$$\Psi^{\text{LAS}}(\boldsymbol{y}) = \{\boldsymbol{b} \in \Omega_K | \boldsymbol{b} \otimes (\mathbf{A}\boldsymbol{y} - \mathbf{H}\boldsymbol{b}) \geq -\boldsymbol{t}^*\} \tag{28}$$

where $\boldsymbol{t}^* = (t_1^*, \ldots, t_K^*)^T$, and multiplication $\otimes$ and inequality $\geq$ are elementwise.

The SLAS, PLAS, and GPLAS detectors all have step-invariant thresholds and, therefore, $t_k^* = t_k$ where t_k is given accordingly in Eqs. (22)–(24).

A WSLAS detector has the same fixed-point region of the SLAS detector as

$$V^{\mathrm{WSLAS}}(\boldsymbol{b}) = \{\boldsymbol{y} \in \mathbb{R}^K | \boldsymbol{b} \otimes [\boldsymbol{y} - (\mathbf{R} - \mathbf{I})\mathbf{A}\boldsymbol{b}] \geq \mathbf{0}\}, \tag{29}$$

which is equal to $V_1^{\mathrm{LML}}(\boldsymbol{b})$, the LML-1 region in Eq. (11). Hence, all the WSLAS detectors are the LML-1 detectors.

The fixed-point region by Eq. (27) can be applied to obtain the BER upper bound [18]. Moreover, Proposition 1 qualitatively indicates the characteristics of the fixed-point region and the relationship between the error performance and the thresholds. First, any bit vector $\boldsymbol{b}$ can be a fixed point with a nonzero probability. Second, given $\boldsymbol{y}$, the GML decision $\boldsymbol{b}^{\mathrm{GML}}(\boldsymbol{y})$ is unique, and the GML decision regions $V^{\mathrm{GML}}(\boldsymbol{b})$ for different $\boldsymbol{b}$'s do not overlap. In contrast, like all the LML-J detectors for $J < K$, a LAS detector may have a set of multiple fixed points $\Psi^{\mathrm{LAS}}(\boldsymbol{y})$, and the fixed-point regions $V^{\mathrm{LAS}}(\boldsymbol{b})$ for different $\boldsymbol{b}$'s may overlap. In the overlapped fixed-point region, one of the fixed points is taken as the demodulated vector depending on the initial vector $\boldsymbol{b}(0)$ and the sequence of $L(n)$ for $n \geq 0$. Third, as the thresholds increase, the fixed-point region expands, the overlapped region expands, and the number of fixed points in the expanded region increases. The error probability increases since the increased fixed points have a lower likelihood. On the other hand, increasing the number of low-likelihood fixed points makes it easier to reach a fixed point, thus decreasing the computational complexity.

The LML-1, WSLAS, LAS, and PLAS detectors have the following duality relationships [18].

i. For any $\boldsymbol{b}$, Proposition 1 implies that

$$V_1^{\mathrm{LML}}(\boldsymbol{b}) = V^{\mathrm{WSLAS}}(\boldsymbol{b}) \subseteq V^{\mathrm{LAS}}(\boldsymbol{b}) \subseteq V^{\mathrm{PLAS}}(\boldsymbol{b}). \tag{30}$$

All the equalities in Eq. (30) hold iff $\mathbf{R} = \mathbf{I}$ where all the GML, LML-J, and LAS detectors are collapsed into the K parallel single-bit MF detectors.

ii. For any $\boldsymbol{y} \in \mathbb{R}^K$, Proposition 1 implies that

$$\Psi_1^{\mathrm{LML}}(\boldsymbol{y}) = \Psi^{\mathrm{WSLAS}}(\boldsymbol{y}) \subseteq \Psi^{\mathrm{LAS}}(\boldsymbol{y}) \subseteq \Psi^{\mathrm{PLAS}}(\boldsymbol{y}). \tag{31}$$

The equality in Eq. (31) can be true for some $\boldsymbol{y}$ even when $\mathbf{R} \neq \mathbf{I}$.

The relationship $V^{\varphi}(\boldsymbol{b}) \subseteq V^{\phi}(\boldsymbol{b})$ in Eq. (30) means that the fixed-point region of the detector φ is inside that of ϕ, which implies the relationship $\Psi^{\varphi}(\boldsymbol{y}) \subseteq \Psi^{\phi}(\boldsymbol{y})$ in Eq. (31) so that a solution of φ is also a solution of ϕ. ϕ performs worse than φ only in the region $V^{\phi}(\boldsymbol{b}) \backslash V^{\varphi}(\boldsymbol{b})$. The error probabilities of these special LAS detectors have the following relationship

$$P_e(\varphi_1^{\mathrm{LML}}) \cong P_e(\varphi^{\mathrm{WSLAS}}) < P_e(\varphi^{\mathrm{LAS}}) < P_e(\varphi^{\mathrm{PLAS}}). \tag{32}$$

To understand the LML region and the LAS fixed point region, **Figure 1(a)** shows the LML-1 and GML (i.e., LML-2) regions, respectively, in a two-bit channel with $R_{12} = \rho > 0$. The four dots are the noise-free MF-output signals $\mathbf{R}\mathbf{A}\boldsymbol{b}$ for $\boldsymbol{b} \in \{-1, +1\}^2$. The LML-1 region of each vector $\boldsymbol{b} \in \{-1, +1\}^2$ where $\boldsymbol{b}$ is an LML-1 vector is bounded by two lines parallel to the axes. For instance, the LML-1 region of $(+1, -1)$ is bounded by the two lines $y_1 = -\rho A_2$ and $y_2 = \rho A_1$.

The LML-1 regions of $(+1, +1)$ and $(-1, -1)$ are identical to the GML regions where an LML-1 detector performs identically to the GML detector. However, the LML-1 regions of $(+1, -1)$ and $(-1, +1)$ are overlapped in the shaded region where both vectors are LML-1 vectors and one of them is the GML vector depending on $\boldsymbol{y}$. The dotted line across the origin is the boundary of the GML regions of $(+1, -1)$ and $(-1, +1)$. In the shaded region, the vector $(-1, +1)$ is the GML vector above the dashed line, and the vector $(+1, -1)$ is the GML vector below the dashed line. There are infinitely many LML-1 detectors due to the infinitely many $\boldsymbol{y}$ each having two choices on $(+1, -1)$ and $(-1, +1)$. In the shaded region, an LML-1 detector will choose either $(+1, -1)$ or $(-1, +1)$ as the demodulated vector. In contrast, the GML detector will choose only the GML vector as the demodulated vector. Hence, only in the shaded region, an LML-1 detector performs worse than the GML detector.

As an LML-1 detector, a WSLAS detector converges to either $(+1, -1)$ or $(-1, +1)$ depending on the sequence of $L(n)$ for $n \geq 0$ and the initial vector $\boldsymbol{b}(0)$. The BER upper bound in the next section applies to all the LML-1/WSLAS detectors and thus is for the worst LML-1/WLAS detector that always chooses the low-likelihood vector in the overlapped region. In **Figure 1(b)** of the PLAS fixed-point regions, the lightly shaded regions have two fixed points, and the deeply shaded regions have three fixed points. With higher thresholds, the PLAS detector has broader fixed-point regions and performs worse than a WSLAS detector.

By Eqs. (12) and (30), an LML-J region with any neighborhood size J is inside the fixed-point region of any LAS detector; and by Eqs. (9) and (31), an LML-J vector with any neighborhood size is a fixed point of any LAS detector. Hence, a LAS detector does not change an initial vector that is an output of an LML-J detector for any J. In contrast, since their decision region is not inside a fixed-point region of the LAS detector, none of the linear detectors (e.g., MF, zero-forcing (ZF), decorrelating, and MMSE detectors), SIC, PIC, DDF, and MMSE-DF detectors [1] can be a fixed point of any LAS detector with probability one. By Theorem 2, any LAS detector can reduce the error probabilities of these detectors, and a WSLAS detector can reduce them to local minima with neighborhood size $J = 1$.

The LML-1, WSLAS, SLAS, LAS, and PLAS detectors have the following relationship of computational complexity

$$C_1^{\text{LML}} \cong C^{\text{SLAS}} > C^{\text{WSLAS}} > C^{\text{LAS}} > C^{\text{PLAS}}. \tag{33}$$

In summary, the GML, LML-J for $J = K - 1, \ldots, 2, 1$, WSLAS, LAS, and PLAS detectors have the relationships of decision regions in Eqs. (12) and (30), sets of demodulated vectors in Eqs. (9) and (31), error probabilities in Eqs. (10) and (32), and computational complexities in Eqs. (14) and (33). Their relationships in error probability and computational complexity are shown in **Figure 2**. All these relationships imply that all the detectors in the LML and LAS families are akin to each other and are in the same class of detectors.

3.6 Upper bounds on BER and lower bound on AME

Like other nonlinear detectors, the BER performance of the LAS detectors is difficult to analyze for finite-size MIMO channels. However, an upper bound of LAS

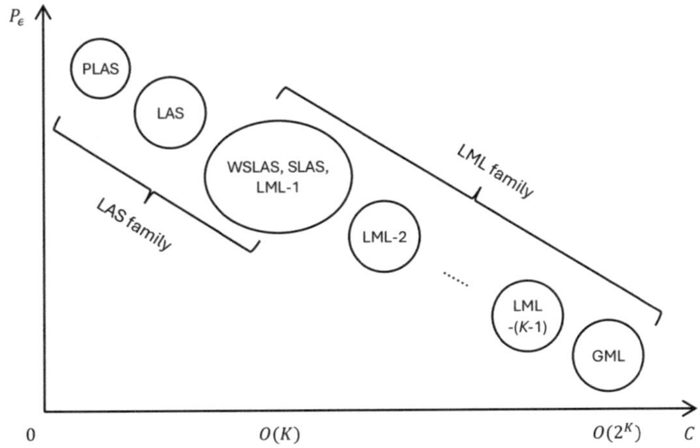

Figure 2.
The relationships of the LML and LAS detectors in terms of error probability P_e and computational complexity C.

BER can be obtained [18], which is comparable to that of the GML detector [19]. Let F_k be the set of indecomposable error vectors affecting bit k and $w(\boldsymbol{\varepsilon})$ be the weight of the error vector $\boldsymbol{\varepsilon} = 0.5(\boldsymbol{b} - \boldsymbol{b}^*)$ where $\boldsymbol{b}$ is the transmitted vector and $\boldsymbol{b}^*$ is the erroneous vector. Let $Q(x) = (2\pi)^{-0.5}\int_x^\infty \exp(-t^2/2)dt$. The following theorem is proved in [18].

Theorem 4: Given any initial detector $\boldsymbol{b}(0)$, the BER of the kth bit for the LAS detector associated with a sequence of $L(n)$ for $n \geq 0$ is upper bounded by

$$P_k^{\text{LAS}}(\sigma) \leq \sum_{\boldsymbol{\varepsilon} \in F_k} 2^{-w(\boldsymbol{\varepsilon})} Q\left(\frac{\boldsymbol{\varepsilon}^T(2\mathbf{H} - \mathbf{T})\boldsymbol{\varepsilon}}{\sigma\sqrt{\boldsymbol{\varepsilon}^T\boldsymbol{\varepsilon}}}\right) \tag{34}$$

where $\mathbf{T} = \text{diag}(t_1^*, \ldots, t_K^*)$ with t_k^* given in Proposition 1.

Confirming the observation on the fixed-point region in Eq. (27), a decrease in the thresholds can decrease the upper bound. Therefore, three changes can result in a decrease in the upper bound: (i) a decrease in the total number of the update candidates, (ii) a decrease in the absolute values of the cross-correlations between the update candidates, and (iii) a decrease in the signal powers of the update candidates.

In particular, by letting $t_k^* = A_k^2$ in Eq. (34), the BER upper bound for all the WSLAS/LML-1 detectors are obtained as in the following corollary.

Corollary 1: The BER of the kth bit for the WSLAS/LML-1 detectors are upper bounded by

$$P_k^{\text{LML}-1}(\sigma) \leq \sum_{\boldsymbol{\varepsilon} \in F_k} 2^{-w(\boldsymbol{\varepsilon})} Q\left(\frac{\boldsymbol{\varepsilon}^T(2\mathbf{H} - \mathbf{A}^2)\boldsymbol{\varepsilon}}{\sigma\sqrt{\boldsymbol{\varepsilon}^T\boldsymbol{\varepsilon}}}\right). \tag{35}$$

It is obvious that the WSLAS/LML-1 detectors achieve the least upper bound in the family of LAS detectors. Since any LML-J vector is also an LML-1 vector in terms of Eqs. (9) and (12), the upper bound in Eq. (35) applies to all the LML-J detectors for any J.

The BER upper bound in Eq. (34) of the LAS detector is comparable with the upper bound of the GML detector obtained by Verdú [19]

$$P_k^{\mathrm{GML}}(\sigma) \le \sum_{\boldsymbol{\varepsilon} \in F_k} 2^{-w(\boldsymbol{\varepsilon})} Q\left(\frac{\sqrt{\boldsymbol{\varepsilon}^T \mathbf{H} \boldsymbol{\varepsilon}}}{\sigma}\right). \tag{36}$$

Note that $\boldsymbol{\varepsilon}^T(2\mathbf{H} - \mathbf{T})\boldsymbol{\varepsilon}/\sqrt{\boldsymbol{\varepsilon}^T\boldsymbol{\varepsilon}} \le \boldsymbol{\varepsilon}^T(2\mathbf{H} - \mathbf{A}^2)\boldsymbol{\varepsilon}/\sqrt{\boldsymbol{\varepsilon}^T\boldsymbol{\varepsilon}} \le \sqrt{\boldsymbol{\varepsilon}^T\mathbf{H}\boldsymbol{\varepsilon}}$. The second equality is true iff $w(\boldsymbol{\varepsilon}) = 1$. That is, the LML-1 detector is the GML detector when a transmitted vector $\boldsymbol{b}$ and an erroneous $\boldsymbol{b}^*$ differ by one bit. The upper bounds in Eqs. (34)–(36) indicate the BER difference of the LAS, LML-1, and GML detectors.

For a detector that achieves BER $P_k(A)$ for the kth bit with energy A^2, the multi-bit (or multiuser) efficiency $\eta = Q^{-2}[P_k(A)]\sigma^2/A^2$ measures the efficiency of signal power usage by the detector in comparison with the MF detector in the single-bit channel with bit energy A^2. The asymptotic multi-bit efficiency (AME) is the limit of η as the noise vanishes $\sigma \to 0$, and thus measures the efficiency of signal power usage in the high SNR regime [19]. By Eq. (34), the following corollary is obtained [18].

Corollary 2: The AME of the LAS detector is lower bounded by

$$\eta_k^{\mathrm{LAS}}(\sigma) \ge \min_{\boldsymbol{\varepsilon} \in F_k}{}^2 \left\{\frac{\left[\boldsymbol{\varepsilon}^T(2\mathbf{H} - \mathbf{T})\boldsymbol{\varepsilon}\right]^+}{A_k\sqrt{\boldsymbol{\varepsilon}^T\boldsymbol{\varepsilon}}}\right\}; \tag{37}$$

particularly the AME for all the WSLAS/LML-1 detectors is lower bounded by

$$\eta_k^{\mathrm{LAS}}(\sigma) \ge \min_{\boldsymbol{\varepsilon} \in F_k}{}^2 \left\{\frac{\left[\boldsymbol{\varepsilon}^T(2\mathbf{H} - \mathbf{A}^2)\boldsymbol{\varepsilon}\right]^+}{A_k\sqrt{\boldsymbol{\varepsilon}^T\boldsymbol{\varepsilon}}}\right\} \tag{38}$$

where $[z]^+ = \max\{0, z\}$.

Unit AME is particularly interesting. In the high SNR regime, BER is dominated by the minimum distance from the transmitted signal to the decision regions of error signals. If the AME for bit k is unit, the minimum distance is determined by the single error, and then bit k's BER achieves the single bit bound $Q(A_k/\sigma)$ as if there was no interfering bit. It follows from Eq. (36) that the GML detector achieves unit AME for bit k in the channels such that [19]

$$d^{\mathrm{GML}}(\boldsymbol{\varepsilon}) \equiv \sqrt{\boldsymbol{\varepsilon}^T\mathbf{H}\boldsymbol{\varepsilon}} \ge A_k, \forall \boldsymbol{\varepsilon} \in F_k. \tag{39}$$

However, none of the well-known suboptimum detectors is known to achieve unit AME except in some trivial two-bit channels and the orthogonal K-bit channel [19]. In contrast, like the GML detector, the LAS detector achieves the following result [18].

Corollary 3: If for each $\boldsymbol{\varepsilon} \in F_k$ with each $w(\boldsymbol{\varepsilon}) = m \le K$ where K can be finite or tend to infinity,

$$d^{\mathrm{LAS}}(\boldsymbol{\varepsilon}) \equiv \frac{\boldsymbol{\varepsilon}^T(2\mathbf{H} - \mathbf{T})\boldsymbol{\varepsilon}}{\sqrt{\boldsymbol{\varepsilon}^T\boldsymbol{\varepsilon}}} \ge A_k, \tag{40}$$

then the LAS detector achieves unit AME for bit k. Furthermore, if Eq. (40) holds for all bits, then the LAS detector achieves unit AME for all bits. In particular, these are all true for the WSLAS/LML-1 detectors if

$$d^{\mathrm{LML-1}}(\boldsymbol{\varepsilon}) \equiv \frac{\boldsymbol{\varepsilon}^T(2\mathbf{H} - \mathbf{A}^2)\boldsymbol{\varepsilon}}{\sqrt{\boldsymbol{\varepsilon}^T\boldsymbol{\varepsilon}}} \ge A_k. \tag{41}$$

The performance of GML, LAS, and LML-1 detectors are determined by their signal distance from the transmitted signal to the decision region of an error signal $d^{\mathrm{GML}}(\boldsymbol{\varepsilon})$, $d^{\mathrm{LAS}}(\boldsymbol{\varepsilon})$, and $d^{\mathrm{LML-1}}(\boldsymbol{\varepsilon})$. The larger the signal distance, the lower the error probability. For the channels that satisfy the condition of Eq. (40) (or (41)), the BER of the LAS (or WSLAS/LML-1) detector in the high SNR region is dominated by the single-bit error signals and approaches the single-bit bound asymptotically. There exist many channels that satisfy Eqs. (40) and (41) regardless of K [18].

4. The WSLAS/LML-1 detectors approaching optimum performance in large MIMO channels

Among all the detectors in the LML and LAS families, the WSLAS/LML-1 detector is most interesting. First, with a per-bit computational complexity linear in K, an LML-1 detector is the simplest among all LML-J detectors for any neighborhood size J. Second, an LML-1 detector can be realized by a WSLAS detector with efficient computations [11–14, 21–25]. Third, a WSLAS/LML-1 detector is the optimum detector in the LAS family. Finally, as presented below, in large MIMO channels, the WSLAS/LML-1 detectors can approach the GML performance in a wide range of channel loads and SNR [21–25].

In a general large MIMO channel, $K \to \infty$ and $N \to \infty$ and the channel load $\alpha = K/N$ is fixed in the large-system limit. The kth channel vector is $\boldsymbol{s}_k = (s_{1k}, \dots, s_{Nk})^T/N^{0.5}$ where s_{ik}'s are i.i.d. with zero mean and unit variance. Therefore, in the large-system limit, the channel vectors converge to the unit length, and the crosscorrelation between two channel vectors converges to zero almost surely, that is, for $j \neq k$

$$\lim_{N\to\infty} \|\boldsymbol{s}_k\| = 1, \lim_{N\to\infty} \boldsymbol{s}_j^H \boldsymbol{s}_k = 0, \text{a.s.} \tag{42}$$

where H is replaced by T as the BPSK signal is considered throughout. The empirical distribution of symbol energies A_k^2 converges to a distribution function. The ratio of neighborhood size J to the number of bits K is denoted by $\beta = J/K$ where J can also grow with K. As the system size tends to infinity, $\beta \to 0$ for any LML-J detector for a fixed J, and $\beta = 1$ for the GML detector.

The general large MIMO channel model with the property of Eq. (42) includes the massive antenna MIMO channels [27], the large random spreading (LRS) CDMA channels with long and short sequences with/without spread spectrum and/or time [21, 22], the LRS CDMA channels with sparse sequences [23, 24], and any large MIMO channel that employs a combination of the multiantenna, CDMA, spectral-temporal spreading sequences, and sparse sequence techniques.

The most important property of the large MIMO channel is Eq. (42). As the channel dimension increases $N \to \infty$, the crosscorrelation matrix $\mathbf{R}$ becomes more and more like an identity matrix $\mathbf{I}$, and the channel becomes more and more like an orthogonal channel. Consequently, the overlapped LML region of multiple vectors in Eq. (12) becomes smaller and smaller, the LML vectors in Eq. (9) become fewer and fewer, and the error probability of an LML detector becomes closer and closer to that of the GML detector. In the large-system limit as $N \to \infty$, the large MIMO channel presents several properties as follows.

4.1 LML characteristic of large MIMO channels

The LML characteristic is one of the most important properties of the large MIMO channels. It is proved for the LRS CDMA channels in Refs. [21, 22] and here is extended to the general large MIMO channels as the theorem below. Let E denote the set of error vectors $\boldsymbol{\varepsilon}$'s, and $I(\boldsymbol{\varepsilon})$ and $w(\boldsymbol{\varepsilon})$ the index set of nonzero elements and the weight of the error vector $\boldsymbol{\varepsilon}$, respectively.

Theorem 5: In the large MIMO channel with any $\alpha > 0$, (i) given any positive integers $M_1 < M_2$, $d^{\mathrm{GML}}(\boldsymbol{\varepsilon}_1) < d^{\mathrm{GML}}(\boldsymbol{\varepsilon}_2)$ a.s. for any $\boldsymbol{\varepsilon}_1, \boldsymbol{\varepsilon}_2 \in E$ such that $I(\boldsymbol{\varepsilon}_1) \subset I(\boldsymbol{\varepsilon}_2)$ and $w(\boldsymbol{\varepsilon}_1) \le M_1 < w(\boldsymbol{\varepsilon}_2) \le M_2$; (ii) for any $\delta > 0$, there exists $M \ge 2$ such that $d^{\mathrm{GML}}(\boldsymbol{\varepsilon}) > \delta$ a.s. for any $\boldsymbol{\varepsilon} \in E$ with $w(\varepsilon) \ge M$.

As indicated in Refs. [21, 22], the LML characteristic means that in the large MIMO channel limit, "the mapping of $\mathbf{SA}$ from Ω_K to $\{\mathbf{SA}\boldsymbol{b} : \boldsymbol{b} \in \Omega_K\} \subset \mathbb{R}^N$ retain the local topology of Ω_K at any bit vector. Standing at the transmitted signal in the $\boldsymbol{r}$ space, one would typically see that the error signals with larger error weights are farther away, and all the error signals with the error weights tending to infinity are infinitely far away. Since the GML decision is based on the nearest distance from $\boldsymbol{r}$ to a signal $\mathbf{SA}\boldsymbol{b}$, the GML BER in the high SNR regime is dominated by the signals that have a one-bit error. This suggests that to achieve the GML detection, one would, without the exhaustive search over the entire set Ω_K, perform only an LML detection."

4.2 Achievability of GML performance by WSLAS/LML-1 detectors

The LML characteristic of large MIMO channels enables the WSLAS/LML-1 detectors to achieve the GML performance in a range of channel loads and SNR. The following theorems are proved for LRS CDMA channels [21, 22], and here are extended to the general large MIMO channels. The critical channel load is equal to $\alpha^* \equiv 1/2 - 1/(4\ln 2)$.

Theorem 6: In the large MIMO channel with $\alpha < \alpha^*$, the AMEs of all the LML detectors converge a.s. to one.

In the large MIMO channel with $\alpha < \alpha^*$, Theorem 6 indicates that the AMEs of the WSLAS/LML-1 detectors converge to one and, therefore, their BERs approach to the single-bit BER bound in the high SNR regime.

Theorem 7: In the large MIMO channel where $\alpha < \alpha^*$ and $\sqrt{N}\sigma = c \in (0, \infty)$ fixed, an LML-1 point is a.s. the GML point.

As pointed out in Refs. [21, 22], in the condition of Theorem 7, "there is typically no LML but the GML point. The likelihood function is sufficiently smooth. Thus, the linear-complex WSLAS detector (and other detectors) with local search can reach the GML point that is usually NP-hard to obtain. All the results are also applicable to the LML detectors with neighborhood sizes greater than one [15–17]."

Although the results in Theorem 6 and Theorem 7 hold with channel load $\alpha < \alpha^* \cong 0.1393$, the simulation results demonstrate that in random spreading CDMA and equal bit powers, the WSLAS detector with a linear per-bit complexity approaches the GML BER in all SNR with channel load as high as 1.05 bits/s/Hz where the dimension is second times Hz [21, 22].

4.3 BER of WSLAS/LML-1 detectors

The BER of the GML detector in the LRS CDMA channel is obtained by the replica analysis by Tanaka [65] and Guo and Verdú [66]. Similarly, we obtain [25] the BER of the WSLAS/LML-1 detectors in the general massive MIMO channels by the replica analysis and by applying the method of Gaussian approximation to the boundary of the fixed-point regions in Eq. (29). The results are presented in the propositions below and can be presented in theorems when the replica method is mathematically proved.

Proposition 2: In the large MIMO channel, for a bit with energy A^2, the BER of WSLAS/LML-1 detectors is the solution of the fixed-point equation

$$p_e(A) = Q\left(\frac{A}{\sqrt{\sigma^2 + 4\alpha \mathbf{E}_A\left(A^2 p_e(A)\right)}}\right) \tag{43}$$

where $\mathbf{E}_A$ is the expectation with respect to A. $p_e(A)$ is also one of the BER solutions of the GML detector and all LML-J detectors for $J \geq 2$.

Proposition 2 points out that the WSLA/LML-1 detectors with linear per-bit complexity can achieve the BER of the GML detector with NP-hard complexity in the large MIMO channels. It also explains why the family of LAS detectors and other detectors with local search can approach the GML performance in the massive antenna MIMO channels [26–58].

As shown in **Figures 3** and **4** [25], the LML-1 BER in Eq. (43) is numerically evaluated and compared with the GML BER [65, 66] in an equal-energy distribution of $A_k = A, \forall k$, under various channel loads α and SNR E_b/N_0. The BER of a SLAS detector [21, 22] in simulation is also obtained with a random initial vector. The single-bit bound (SBB) on BER is also shown. Several results can be observed in the figures.

First, like the GML BER [65], as α becomes larger, multiple LML-BERs coexist.

Second, as shown in **Figure 3**, the AT line divides the region of channel load and SNR $(\alpha, E_b/N_0)$ into two regions. In region A, the LML-1 and GML detectors perform

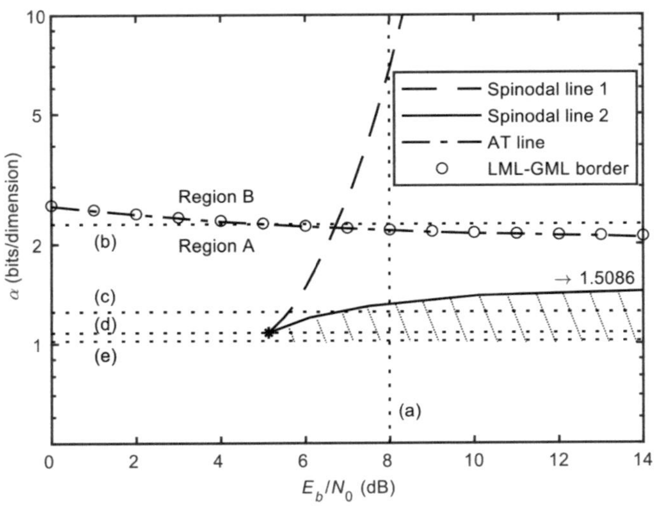

Figure 3.
*A practical region of α and E_b/N_0. Lines (a)–(e) indicate α and E_b/N_0 used in **Figure 4(a)–(e)**, respectively. The spinodal lines 1, 2 intersect at *: $(\alpha, E_b/N_0) = (1.08, 5.13)$ (bits/dimension, dB).*

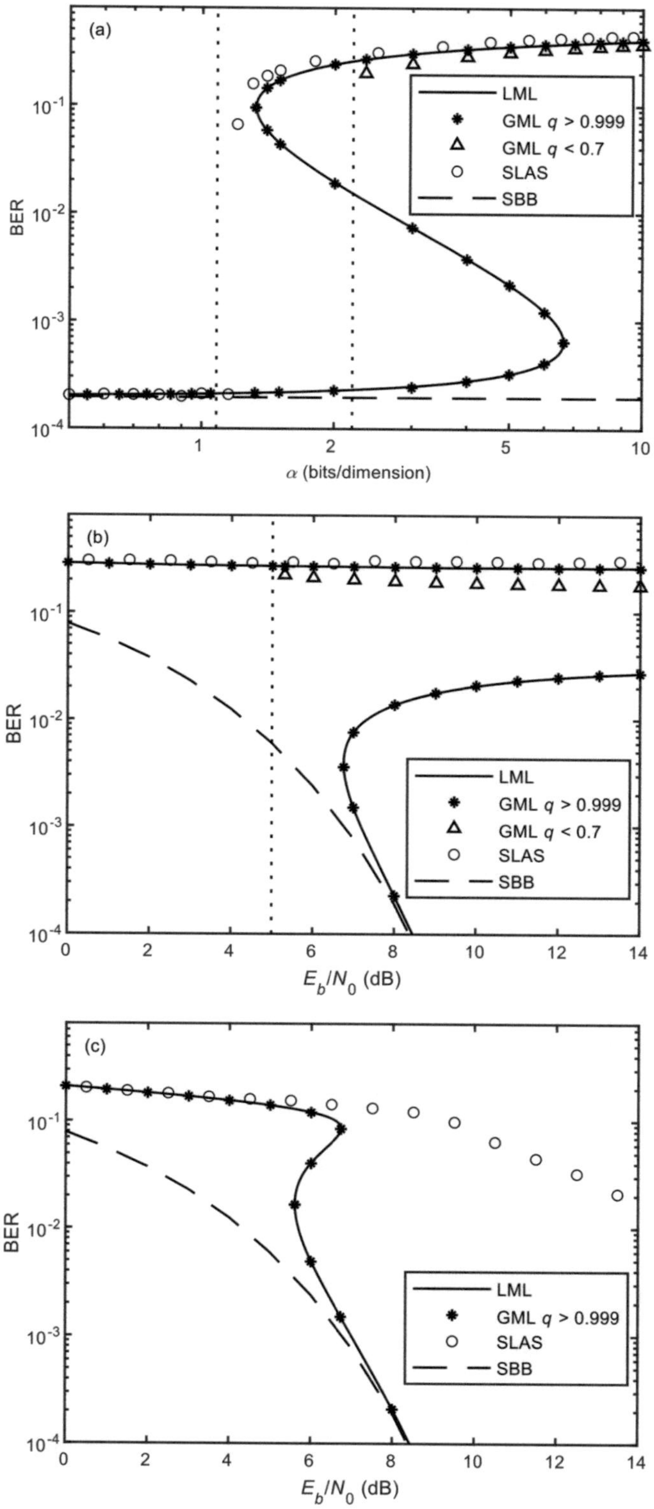
(a)
LML
GML q > 0.999
GML q < 0.7
SLAS
SBB
BER
α (bits/dimension)
(b)
LML
GML q > 0.999
GML q < 0.7
SLAS
SBB
BER
Eb/N0 (dB)
(c)
LML
GML q > 0.999
SLAS
SBB
BER
Eb/N0 (dB)

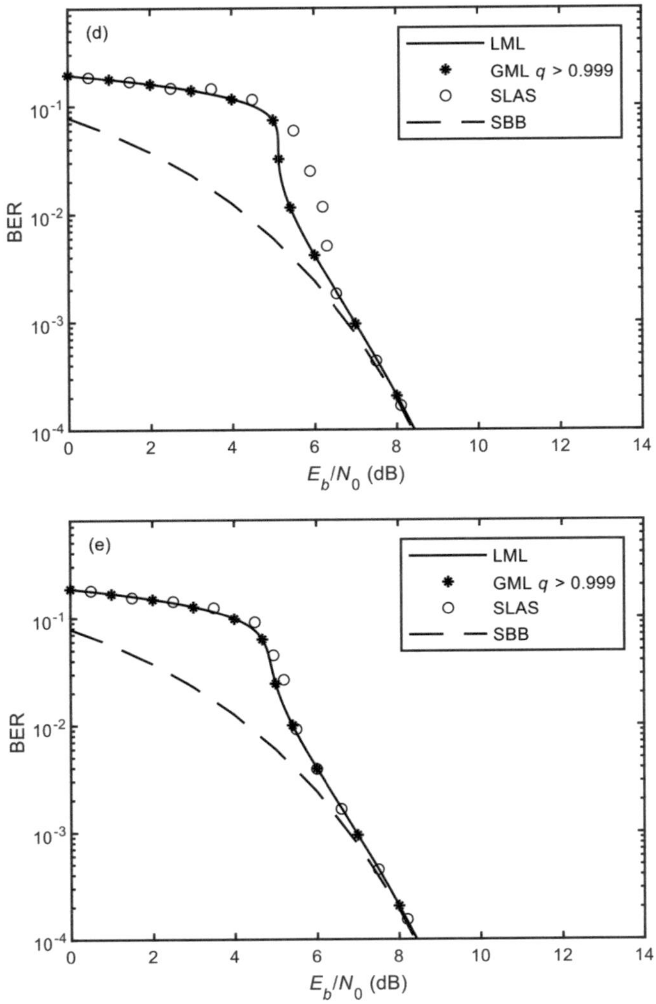

Figure 4.
(a) BER versus α with $E_b/N_o = 7.99$ dB. The left vertical line is at $\alpha = 1.08$ bits/dimension, and the right at $\alpha = 2.20$ bits/dimension which separates regions A and B. (b) BER versus E_b/N_o with $\alpha = 2.30$ bits/dimension. The vertical line indicates $E_b/N_o = 5.0$ dB that separates regions A and B. (c) BER versus E_b/N_o with $\alpha = 1.25$ bits/dimension. (d) BER versus E_b/N_o with $\alpha = 1.08$ bits/dimension. (e) BER versus E_b/N_o with $\alpha = 1.02$ bits/dimension. The values of $(\alpha, E_b/N_o)$ in (c)–(e) all are in region A; therefore, the gap between the LML and GML BERs is small, and so is the gap between their multiuser efficiencies.

identically with the only saddle-point solution of $q = 1$. In Region B, the GML detector has two bad solutions with $q < 0.7$ and $q = 1$, respectively. The LML-1 detector performs worse than the GML detector only for the additional bad solution of the high BER with $q < 0.7$, which the LML-1 detector cannot achieve.

Third, it is particularly practically interesting that there is a region below the spinodal line 2 denoted by the sloped dotted lines in **Figure 3**. In this region, the LML-1 BER is identical to the GML BER, and the channel load is greater than 1 bit/dimension. Specifically, for $\alpha \in [1, 1.08)$ bit/dimension, the SNR can be any $E_b/N_0 \in (-\infty, \infty)$. On the other hand, in the high SNR region as $E_b/N_0 \to \infty$, the channel load can be as high as $\alpha \cong 1.5086$ bits/dimension.

Fourth, for $\alpha \in [1, 1.08)$ bit/dimension, the practical SLAS detector (an LML-1 detector) in simulation can achieve the GML/LML-1 BER in the large-system limit. It implies that the linear-complex WSLAS detector can approach the GML performance in massive antenna MIMO systems. More discussions of **Figure 4** are presented in Section 4.5.

In summary, in large MIMO channels, the linear-complex WSLAS/LML-1 detectors can achieve all the good BER solutions of the NP-hard GML detector.

4.4 Achievable channel load

By means of Eq. (43), the multi-bit efficiency of the LML-1 detector is the solution to the following fixed-point equation $\eta = \sigma^2/\{\sigma^2 + 4\alpha \mathbf{E}_A[A^2 p_e(A)]\}$ and the AME is

$$\hat{\eta} = \lim_{\sigma \to 0} \frac{\sigma^2}{\sigma^2 + 4\alpha \mathbf{E}_A[A^2 p_e(A)]}. \tag{44}$$

Since $p_e(A)$ can be multiple-valued, and so are the multi-bit efficiency and AME. In the large-system limit, the BER for the bit of energy A^2 is $p_e(A)$. If the AME equals one, in the high SNR regime, the LML-1 BER $p_e(A)$ can approach the SBB $p_s(A) = Q(A/\sigma)$ as if there was no interference bit.

The cutoff channel load (CCL) in the high SNR regime is defined as

$$\alpha^* = \sup_{\alpha > 0}\{\alpha : \hat{\eta} = 1\}.$$

The CCL α^* is the marginal channel load at which the bad solution disappears in the high SNR regime. All channel loads $\alpha < \alpha^*$ are asymptotically achievable, under which all bits can achieve the SBB BER in the high SNR regime. On the other hand, all channel loads $\alpha > \alpha^*$ are not achievable.

To obtain the CCL for the LML-1 detector in an equal-energy distribution, we have the following lemma [25].

Lemma 1: For the fixed-point equation of p

$$p = Q\left(\frac{1}{\sqrt{4\alpha p}}\right), \alpha > 0, \tag{45}$$

there exist $\alpha_0 \cong 1.5086$ and $p_0 \cong 0.1169$ such that (i) if $\alpha < \alpha_0$, only $p = 0$ solves Eq. (45); (ii) if $\alpha = \alpha_0$, then both $p = p_0$ and $p = 0$ solves Eq. (45); (iii) if $\alpha > \alpha_0$, in addition to $p = 0$ there exist two points $p_1 \in (0, p_0)$ and $p_2 \in (p_0, 1)$ that solve Eq. (45).

By Eq. (43), as $\sigma \to 0$, the LML-1 BER in the equal-energy distribution is a fixed point that solves Eq. (45). In terms of Lemma 1, the LML-1 CCL is $\alpha^* = \alpha_0 \cong 1.5086$ bits/dimension. α_0 is the marginal channel load such that the bad solution disappears in the high SNR regime. Consequently, the LML-1 and GML detectors have the same CCL.

It is interesting that the LML-1/GML CCL α_0 is greater than 1 bit/dimension. In contrast, the channel load equals 1 bit/dimension for a set of orthogonal channel vectors and for the TDMA/FDMA systems where both transmitter and receiver have a single antenna. Moreover, the AMEs of the MMSE, decorrelating and MF detectors all are smaller than one [67] and, therefore, their CCLs are equal to zero.

The LML-1/GML CCL is low in the equal-energy distribution and can be arbitrarily high in an unequal-energy distribution. To obtain this, the result of LML-1 CCL is extended to an arbitrary energy distribution [25].

Proposition 3: The LML-1 CCL in an arbitrary energy distribution is equal to

$$\alpha^* = \frac{I}{4\mathbf{E}_A\left[A^2 Q\left(A/I^{1/2}\right)\right]} \tag{46}$$

where the interference energy at α^* satisfies

$$I = \frac{\mathbf{E}_A^2\left[A^3 \exp\left(-A^2/(2I)\right)\right]}{8\pi\mathbf{E}_A^2\left[A^2 Q\left(A/I^{1/2}\right)\right]}. \tag{47}$$

To obtain α^* numerically in an arbitrary energy distribution, we can first iteratively solve I from Eq. (47) and then obtain α^* from Eq. (46). Letting $A = 1$, Eqs. (46) and (47) can be applied to determine iteratively the LML-1 CCL α_0 in the equal-energy distribution as given in Lemma 1.

As shown in **Figure 5** for a two-class system, an unequal energy distribution can significantly increase the region of achievable channel load. As the energy difference between the two classes increases, the CCL monotonically increases and can be arbitrarily close to α_0/λ_1 if A_2 is sufficiently small. This is true in general; that is, without limitation on the energy distribution, any channel load $\alpha > 0$ is achievable for the LML-1 detector. Moreover, $\alpha^* \geq \alpha_0$ where the equality holds iff all bits have equal-energy. Hence, an unequal energy distribution is preferable to an equal-energy distribution.

According to Eq. (43), the LML-1 detector has the total effective multi-bit interference $4\alpha\mathbf{E}_A\left[A^2 p_e(A)\right]$. A bit of energy A^2 interferes with others only through

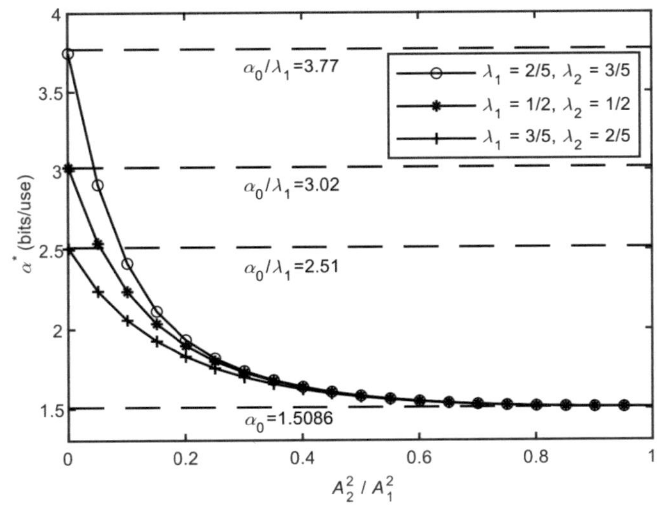

Figure 5.
The cutoff channel load α^* versus energy distribution in the two-class system where the ith class has energy A_i^2 and percentage of the population λ_i, $i = 1, 2$.

$4A^2p_e(A)$. Below the spinodal line 2 of **Figure 3** where the bad solution disappears, the BER $p_e(A)$ decreases exponentially fast to zero as the energy increases; therefore, a strong bit yields a weak interference. In particular, a bit with $A \to \infty$ does not interfere with others since $A^2p_e(A) \to 0$ and thus the LML-1 detector is effective in near-far resistance defined in [19]. Moreover, a weak bit also yields a weak interference. Consequently, the overall interference is low. This is why the LML-1 detector is superior to all linear detectors. For example, the linear MMSE detector has the multi-bit efficiency $\eta = \sigma^2/\{\sigma^2 + \alpha \mathbf{E}_A[A^2/(1+\eta A^2\sigma^{-2})]\}$ [67]. A bit of energy A^2 interferes others through $A^2/(1+\eta A^2\sigma^{-2})$, which does not vanish as energy increases; therefore, the overall interference is high.

4.5 Practical achievability of GML BER

As pointed out in the preceding subsection, there is a special region of channel load and SNR as indicated by the sloped dotted lines in **Figure 3**. When the channel load α is lower than the spinodal line 2 where the bad solution disappears, there is only a good solution, and the LML-1 and GML detectors perform identically. Specifically, when $\alpha < 1.08$ bits/dimension, there is no other solution but a good solution for all SNR. If $E_b/N_0 > 5.13$ dB, the channel load can be higher than 1.08 bits/dimension. As SNR tends to infinity, the highest channel load is $\alpha = 1.5086$ bits/dimension. These facts imply the possible achievability of the NP-hard GML BER in a practical large MIMO system by a linear-complex LML-1 detector that uses only the local information of the current bit vector in a search.

To investigate the practicability, the SLAS detector [21, 22] is examined in the simulation. The SLAS detector, which belongs to the WSLAS detectors, is an LML-1 detector and has a linear per-bit computational complexity [21, 22]. In the simulation, the equal-energy system with $K = 8192$ bits is considered. A random bit vector is used for the SLAS detector to initiate the likelihood ascent search. The BER is obtained by averaging over a number of transmissions such that at least a total of 300 bit errors occur.

In **Figure 4(a)** with $E_b/N_0 = 7.99$ dB. The SLAS detector behaves differently in three regimes of channel load. In the regime of $\alpha < 1.15$ bits/dimension where there is only the good LML-1/GML solution, the SLAS detector approaches the BER that is close to the SBB. In the regime between the spinodal line 2 and the AT line with $1.15 < \alpha < 2.20$ bits/dimension where one good and one bad LML-1/GML solutions coexist, the SLAS detector can only approach the bad BER. On the other hand, in the regime above the AT line (region B) with $\alpha > 2.20$ bits/dimension where the good LML-1/GML solution, the bad LML-1/GML solution and the bad GML solution coexist, the SLAS detector can only approach the higher LML-1/GML BER of the bad solution. This confirms the preceding result that the lower GML BER of the bad solution with $q < 0.7$ cannot be approached by an LML-1 detector.

In **Figure 4(b)** with $\alpha = 2.3$ bits/dimension, the left side of the vertical line with $E_b/N_0 = 5.0$ dB is in region A where there is only one bad LML-1/GML solution, and the right side is in region B where an additional bad GML solution appears. However, as expected, in the entire considered region, the SLAS detector can only approach the bad LML-1/GML solution with the higher BER.

In **Figure 4(c)** with $\alpha = 1.25$ bits/dimension, the system parameters $(\alpha, E_b/N_0)$ are in region A below the AT line, and therefore, the GML and LML-1 detectors perform identically. For $E_b/N_0 < 6.73$ dB above the spinodal line 2, the bad LML-1/GML solution exists, and the SLAS detector can only approach the bad solution. For

$E_b/N_0 > 6.73$ dB below the spinodal line 2, only the good LML-1/GML solution exists. However, since $\alpha = 1.25$ bits/dimension is too close to the spinodal line 2, the SLAS detector approaches a BER with a big gap to the analytical solution. The reason is that when the system parameters $(\alpha, E_b/N_0)$ are close to the spinodal line 2, the SLAS performance is sensitive to the system size. It is observed that the BER gap decreases as the number of bits increases. That is, the number of bits $K = 8192$ is not sufficiently large for the SLAS detector to approach the analytical solution.

In **Figure 4(d)** with $\alpha = 1.08$ bits/dimension that is farther away from the spinodal line 2, the SLAS detector can approach the good LML-1/GML BER but is slightly affected by the spinodal line 2 when E_b/N_0 is close to 5.13 dB at which the two spinodal lines intersect.

In **Figure 4(e)** with $\alpha = 1.02$ bits/dimension where the system parameters $(\alpha, E_b/N_0)$ are far away from the spinodal line 2, the SLAS detector can approach the unique good LML-1/GML BER.

These simulation results demonstrate that the linear-complex SLAS detector using only the local information in the likelihood ascent search can approach the NP-hard GML performance in a practical large MIMO system. This is theoretically significant and practically useful when the LAS is rolled out and implemented on hardware suitable for fast parallel computations, as presented in the next section.

The numerical results demonstrate that in the large-system limit, the BER difference between the LML with $\beta \to 0$ and the GML with $\beta = 1$ is small. The BER difference among all LML-J detectors with any $\beta = J/K \in [0, 1]$ must be small. This implies that in a sufficiently large practical MIMO system, the BER difference among all LML-J detectors for $J > 1$ must be small. Since the SLAS/LML-1 detector in a system of $K = 8192$ bits can already approach the BER with a small gap to the analytical GML BER, any other LML-J detector for $J > 1$ must not considerably improve the BER. On the other hand, the computational complexity of an LML-J detector grows polynomially fast with J [15–17]. Hence, in a practical, sufficiently large MIMO system, it is preferred to employ the WSLAS/LML-1 detector to achieve the best tradeoff between the BER performance and computational complexity.

5. Application to next generation wireless networks

The next generation 6G and beyond wireless networks is expected to connect a large number of wireless devices and push the spectral load to its limit [59, 60]. On one hand, massive MIMO antennas can significantly increase the dimension of information transmission in a massive MIMO system. On the other hand, it is critical to exploit the spectral efficiency at the physical layer by an advanced detector. The family of LML detectors and the family of LAS detectors are particularly suitable for the symbol detector in a massive MIMO system to achieve optimum spectral efficiency. Among all the LML and LAS detectors the WSLAS/LML-1 detector can achieve GML detection while the per-bit complexity is linear and, therefore, is the best candidate for the 6G and beyond wireless networks. Aiming at this objective, several practical issues are addressed below.

5.1 Implementation of WSLAS detectors on a multilayer perceptron

Like all the LMLAS and LAS detectors, the WSLAS detector searches a bit vector in iteration. Iterative computation might be the major factor that hinders the WSLAS

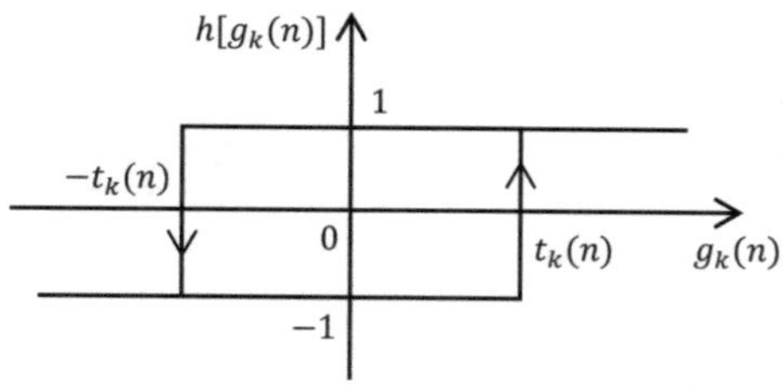

Figure 6.
The output function of the kth neuron in the nth layer $b_k(n+1) = h[g_k(n)]$. The arrows indicate the directions of state transitions. The neural state does not change when $g_k(n) \in [-t_k(n), t_k(n)]$.

detector from being applied in practical MIMO systems where symbols must be detected in real-time. To address the issue of iterations, we propose to roll out the iterations of the WSLAS detector.

Specifically, consider the neural network architecture of a multilayer perceptron. The search step n indexes a layer. Each layer has K neurons. The output state of the kth neuron in the nth layer is $b_k(n) \in \{-1, 1\}$. Two adjacent layers are connected by the matrix $-\mathbf{H}$. According to Eq. (16), the kth neuron in the nth layer has the input of $-\sum_{i=1}^{K} H_{ki} b_i(n-1)$, the bias of $A_k y_k$, the neural threshold $t_k(n)$ calculated by Eq. (21), and by Eq. (20), the neural output function in **Figure 6**. The gradient in a layer is updated from the previous layer according to Eq. (18).

The WSLAS detector takes n^* search steps to reach a fixed point, and $n^* = cK$ is linear in K, which is also the per-bit complexity in terms of the number of additions used per bit. Thus, the number of layers in the multilayer perceptron is equal to cK where c is much smaller than one. For example, the SLAS detector updates one bit in each step and c is equal to the bit flip rate (BFR). In the simulations [21, 22], the number of bits is $1136 \le K \le 3328$, the channel load and SNR are in a wide practical range, and the MF detector is employed as the initial detector. The SLAS detector approaches the GML BER in the large-system limit while the BFR is $0.01 < c < 0.49$. In the general case of $K = 3000$ and channel load $\alpha = 0.7$ bits/dimension, the BFR is $c = 0.2$ and then the number of layers is about $cK = 600$. In the worst case of $K = 3328$ and channel load $\alpha = 1.2$ bits/dimension, the BFR is $c = 0.49$ and the number of layers is about $cK = 1630$, a deep multilayer perceptron.

By implementing the WSLAS detector on the architecture of a multilayer perceptron, the receiver can demodulate the transmitted symbols one vector after another in the pipeline and in real-time. The only cost of time incurred is the latency in the computation time over the cK layers. Hence, the rollout of the WSLAS detector on a multilayer perceptron can be implemented in practical MIMO systems to achieve optimum performance in symbol detection and spectral efficiency.

5.2 EHE and FMD criteria

The latency of the WSLAS detector on a multilayer perceptron depends on n^*, the number of steps to reach a fixed point. To reduce the latency, three methods can be applied to decrease n^* while guaranteeing convergence to an LML-1 point.

First, the search sequence $L(n), n \ge 0$ can be properly designed. In a few initial steps, more bits are updated in each step with a large $|L(n)|$ in order to converge fast in fewer steps. Gradually fewer bits are updated in each step with a smaller $|L(n)|$. Finally, one bit is updated in each step to converge to an LML-1 vector.

Second, the fastest metric descent (FMD) criterium [11] can be applied. In each step, only the bits with the largest $|g_k(n)| - 0.5A_k$ are updated. Under the FMD criterium, the WSLAS detector can converge to an LML-1 vector in fewer steps and, therefore, the number of layers can be smaller.

Third, the eliminating highest error (EHE) criterium [11] can be employed. In each step, only the bits with the highest $g_k(n)/A_k$ are updated. The higher the relative gradient component $g_k(n)/A_k$, the higher the probability of correct bit flip. Hence, the EHE criterium can enable the WSLAS detector to converge to a fixed point in fewer steps, and then the number of layers can be smaller.

5.3 Random sparse codes

Random sparse codes can be applied to transmitted bits. The kth bit is multiplied by the code vector $\boldsymbol{c}_k = (c_{k1}, \dots, c_{kM})^T$ before transmission. Each code vector $\boldsymbol{c}_k \in \{-L^{-0.5}, 0, L^{-0.5}\}^M$ has L nonzero chips that are randomly located and equiprobably take on $\pm L^{-0.5}$. When L/M is small, the codes are sparse as most of the chips are zero. The advantage of random sparse codes is that only the nonzero chips need transmission power and, therefore, the overall transmission power is much less.

In the simulation [23, 24], the number of bits is $16 \le K \le 2000$, SNR is $E_b/N_0 = 8$ dB and channel load is $\alpha = 0.8$ bit/dimension. The results demonstrate that when there are as few as $L = 16$ nonzero chips, the WSLAS detector can already approach the GML/LML-1 BER with non-sparse random codes. This implies that the WSLAS detector with a set of random sparse codes can significantly reduce transmission power while achieving optimum performance in massive MIMO channels.

5.4 LML-J for small MIMO channels

In a small or middle MIMO channel, the numbers of Tx and Rx antennas are small. The likelihood function is rough and presents more LML points with low likelihood [21, 22]. In general, the WSLAS detector is more likely trapped in an LML-1 point with low likelihood. Consequently, the WSLAS/LML-1 detector often performs far worse than the GML detector. To address the problem, the EHE and FMD criteria [11] can be applied. In addition, other methods can also be considered. For example, a more complex and highly performed MMSE detector can be employed as the initial detector. Another method is to initial the WSLAS detector with multiple random vectors and select the optimum fixed point with the maximum likelihood as the demodulated vector.

In small or middle MIMO channels, the computational complexity of the WSLAS detector in the order of $O(K)$ is small. Then an LML-J detector with $J \ge 2$ can be applied and the per-bit computational complexity is in the order of $O\left(\sum_{i=1}^{J} \binom{K}{i}\right)$. For example, suppose the BFR is $c = 0.5$ in all cases. In a massive MIMO channel where $K_m = 10000$ bits are transmitted, the per-bit computational complexity is $cK_m = 5000$. In a small MIMO channel where $K_s = 100$ bits are transmitted, an LML-2 detector has the per-bit computational complexity about $c(K_s + 0.5K_s^2) = 2550$, in the same order as the WSLAS/LML-1 detector in the massive MIMO channel.

6. Conclusions

The family of local maximum likelihood (LML) detectors consists of all LML-J detectors that each achieve the maximum likelihood in its neighborhood of size J, including the global maximum likelihood (GML) detector that achieves the maximum likelihood in the entire set of bit vectors. The family of likelihood ascent search (LAS) detectors each searches out a sequence of vectors with a monotonical likelihood ascent and, therefore, converges to a fixed point with a finite number of steps. The wide-sense sequential LAS (WSLAS) detectors are particularly interesting. First, they are the LML-1 detectors and belong to both families. Second, their computational complexity per bit equals cK linear in K with $c < 0.5$. Third, they can approach the NP-hard GML detector in large MIMO channels. Fourth, under several criteria, they can converge to an LML-1 vector in fewer steps. Finally, the WSLAS detectors, like all LAS detectors, can be implemented on the architecture of a multilayer perceptron to achieve real-time symbol detection.

The general large MIMO channel is proven to possess the LML characteristic; therefore, a local search detector with likelihood ascent, like the WSLAS detectors, can approach the GML detection. Moreover, in the large-system limit, the LML-1 and GML detectors are proven to perform identically over a practically wide range of channel load and SNR, and the LML-1/GML BER achieves the single-bit BER in the high SNR regime with the channel load as high as up to 1.5086 bits/dimension.

With the capability of exploiting the LML characteristic and the potential high spectral efficiency of massive MIMO channels, the families of LAS and LML detectors are expected to play a role in the next generation 6G and beyond wireless networks.

Acknowledgements

This book chapter was published as a preprint on arXiv with the title "The family of LML detectors and the family of LAS detectors for massive MIMO communications" arXiv:2407.19709 in 2024.

Author details

Yi Sun
Electrical Engineering Department, The City College of City University of New York, New York, NY, USA

*Address all correspondence to: ysun@ccny.cuny.edu

References

[1] Paik JK, Katsaggelos AK. Image restoration using a modified Hopfield network. IEEE Transactions on Image Processing. 1992;**1**(1):49-63

[2] Sun Y, Yu S. A modified Hopfield neural network used in bilevel image restoration and reconstruction. In: Proc. Int. Symp. Inform. Theory and its Appl. Singapore, Singapore: IEEE; 1992

[3] Sun Y, Yu S. An eliminating highest error criterion in Hopfield neural network for bilevel image restoration. In: Proc. Int. Symp. Inform. Theory and its Appl. Singapore, Singapore: IEEE; 1992

[4] Sun Y, Li J-G, Yu S-Y. Improvement on performance of modified Hopfield neural network for image restoration. IEEE Transactions on Image Processing. 1995;**4**(5):688-692

[5] Sun Y. A generalized updating rule for modified Hopfield neural network. In: Proc. Int. Conf. Neural Networks (ICNN'97). Houston, TX, USA: IEEE; 1997

[6] Sun Y. A generalized updating rule for modified Hopfield neural network for quadratic optimization. Neurocomputing. 1998;**19**(1–3):133-143

[7] Sun Y. Hopfield neural network based algorithms for image restoration and reconstruction. I. Algorithms and simulations. IEEE Transactions on Signal Processing. 2000;**48**(7):2105-2118

[8] Sun Y. Hopfield neural network based algorithms for image restoration and reconstruction. II. Performance analysis. IEEE Transactions on Signal Processing. 2000;**48**(7):2119-2131

[9] Hopfield JJ, Tank DW. 'Neural' computation of decisions in optimization problems. Biological Cybernetics. 1985; **52**(3):141-152

[10] Zhou Y-T, Chellappa R, Vaid A, Jenkins BK. Image restoration using a neural network. IEEE Transactions on Acoustics, Speech, and Signal Processing. 1988;**36**(7): 1141-1151

[11] Sun Y. Eliminating-highest-error and fastest-metric-descent criteria and iterative algorithms for bit-synchronous CDMA multiuser detection. In: Proc. IEEE Int. Conf. Commu., ICC'98. Atlanta, Georgia, USA: IEEE; 1998

[12] Sun Y. A generalized search rule of likelihood ascent search detectors for CDMA multiuser detection. In: Proc. 5th Conf. Info. Sys. Analy. & Synth./3rd Conf. Sys., Cybern. And Info., ISAS´99/SCI'99. Orlando, Florida, USA: IIIS (International Institute of Informatics and Systems); 1999

[13] Sun Y. A family of linear complexity likelihood ascent search multiuser detectors for CDMA communications. In: Conf. Record 34th Asilomar Conf. Signals, Systems and Computers. Pacific Grove, CA, USA: IEEE; 2000

[14] Sun Y. A family of linear complexity likelihood ascent search detectors for CDMA multiuser detection. In: 2000 IEEE 6th Int. Symp. Spread Spectrum Tech. and Appl. ISSTA. Parsippany, NJ, USA: IEEE; 2000. p. 2000

[15] Sun Y. A family of likelihood ascent search detectors achieving local maximum likelihood with an arbitrary neighborhood size for CDMA multiuser detection. In: Proc. 38th Ann. Allerton Conf. Commun., Control, and Computing. Monticello, IL, USA: Coordinated Science Laboratory at UIUC (University of Illinois at Urbana-Champaign); 2000

[16] Sun Y. Local maximum likelihood multiuser detection. In: Proc. 34th Ann. Conf. Info. Sci. And Sys., CISS'2001. Princeton, NJ, USA: IEEE; 2001

[17] Sun Y. Local maximum likelihood multiuser detection for CDMA communications. In: Pro. Int. Conf. Info. Techn.: Coding and Computing. Las Vegas, NV, USA: IEEE; 2001

[18] Sun Y. A family of likelihood ascent search multiuser detectors: An upper bound of bit error rate and a lower bound of asymptotic multiuser efficiency. IEEE Transactions on Communications. 2009;**57**(6):1743-1752

[19] Verdú S. Multiuser Detection. New York: Cambridge University Press; 1998

[20] Sun Y. Local maximum likelihood multiuser detection. In: ARL Collab. Tech. Alliance Commu. And Networks-Progress Report 4Q01. Adelphi, MD, USA: ARL; 2001

[21] Sun Y. A family of likelihood ascent search multiuser detectors: approach to single-user performance via quasi-large random sequence CDMA, arXiv preprint, arXiv:0711.3867, 2007

[22] Sun Y. A family of likelihood ascent search multiuser detectors: Approaching optimum performance via random multicodes with linear complexity. IEEE Transactions on Communications. 2009; **57**(8):2215-2220

[23] Sun Y. Quasi-large sparse-sequence CDMA: Approach to single-user bound by linearly-complex LAS detectors. In: 42nd Ann. Conf. Info. Sci. Sys., CISS. Princeton, NJ, USA: IEEE; 2008. p. 2008

[24] Sun Y, Xiao J. Multicode sparse-sequence CDMA: Approach to optimum performance by linearly complex WSLAS detectors. Wireless Personal Communication. 2013;**71**(2):1049-1056

[25] Sun Y, Zheng L, Zhu P, Wang X. On optimality of local maximum-likelihood detectors in large-scale MIMO channels. IEEE Transactions on Wireless Communications. 2016;**15**(10): 7074-7088

[26] Vardhan KV, Mohammed SK, Chockalingam A, Rajan BS. A low-complexity detector for large MIMO systems and multicarrier CDMA systems. IEEE Journal on Selected Areas in Communications. 2008;**26**(3): 473-485

[27] Albreem MA, Juntti M, Shahabuddin S. Massive MIMO detection techniques: A survey. IEEE Communication Surveys and Tutorials. 2019;**21**(4):3109-3132

[28] Mohammed SK, Chockalingam A, Rajan BS. A low-complexity near-ML performance achieving algorithm for large MIMO detection. In: 2008 IEEE Int. Symp. Infor. Theory. Toronto, ON, Canada: IEEE; 2008

[29] Chockalingam A. Low-complexity algorithms for large-MIMO detection. In: 4th Int. Symp. Commu., Control and Signal Proc. (ISCCSP). Limassol, Cyprus: IEEE; 2010

[30] Ducoing JDL, Yi N, Ma Y, Tafazolli R. Using real constellations in fully-and over-loaded large MU-MIMO systems with simple detection. IEEE Wireless Commuication Letters. 2016; **5**(1):92-95

[31] Liu W, Zhang Y, Jin M. Lagrangian detection for generalized space-shift keying MIMO systems. IEEE Transactions on Vehicular Technology. 2017;**66**(9):8585-8589

[32] Challa NR, Bagadi K. Likelihood ascent search detection for coded massive MU-MIMO systems to mitigate IAI and MUI. Radioelectronics and Communications Systems. 2020;**63**(5): 223-234

[33] Li L, Meng W, Li C. Semidefinite further relaxation on likelihood ascent search detection algorithm for high-order modulation in massive MIMO system. IET Communication. 2017;**11**(6): 801-808

[34] Solanki M, Gupta S. Robust conjugate-gradient based LAS detector for massive MIMO systems. International Journal of Electronics. 2022;**109**(5):794-810

[35] Hu J, Song S, Wang Z. A novel low-complexity massive MIMO detector with near-optimum performance. In: IEEE Int. Symp. Circ. Sys. (ISCAS). Singapore, Singapore: IEEE; 2024

[36] Srinidhi N, Mohammed SK, Chockalingam A, Rajan BS. Low-complexity near-ML decoding of large non-orthogonal STBCs using reactive tabu search. In: IEEE Int. Symp. Infor. Theory. Seoul, Korea: IEEE; 2009

[37] Datta T, Srinidhi N, Chockalingam A, Rajan BS. Random-restart reactive tabu search algorithm for detection in large-MIMO systems. IEEE Communication Letters. 2010;**14**(12): 1107-1109

[38] Datta T, Srinidhi N, Chockalingam A, Rajan BS. A hybrid RTS-BP algorithm for improved detection of large-MIMO M-QAM signals. In: Nat. Conf. Commu. (NCC). Bangalore, India: IEEE; 2011

[39] Srinidhi N, Datta T, Chockalingam A, Rajan BS. Layered tabu search algorithm for large-MIMO detection and a lower bound on ML performance. IEEE Transactions on Communications. 2011;**59**(11):2955-2963

[40] Karthikeyan M, Saraswady D. Low complexity layered tabu search detection in large MIMO systems. AEU-International Journal of Electronics and Communications. 2018;**83**:106-113

[41] Nguyen NT, Lee K, Dai H. QR-decomposition-aided tabu search detection for large MIMO systems. IEEE Transactions on Vehicular Technology. 2019;**68**(5):4857-4870

[42] Chakraborty S, Sinha NB, Mitra M. Low complexity, pairwise layered Tabu search for large scale MIMO detection. Wireless Personal Communication. 2023;**128**(3):1689-1713

[43] Sah AK, Chaturvedi AK. An unconstrained likelihood ascent based detection algorithm for large MIMO systems. IEEE Transactions on Wireless Communications. 2017;**16**(4):2262-2273

[44] Sah AK, Chaturvedi AK. Sequential and global likelihood ascent search-based detection in large MIMO systems. IEEE Transactions on Communications. 2018; **66**(2):713-725

[45] Sah AK, Chaturvedi AK. Reduced neighborhood search algorithms for low complexity detection in MIMO systems. In: IEEE Global Commu. Conf. GLOBECOM; San Diego, CA, USA: IEEE; 2015

[46] Mann P, Sah AK, Budhiraja R, Chaturvedi AK. Bit-level reduced neighborhood search for low-complexity detection in large MIMO systems. IEEE Wireless Communications Letters. 2018; **7**(2):146-149

[47] Li P, Murch RD. Multiple output selection-LAS algorithm in large MIMO

systems. IEEE Communications Letters. 2010;**14**(5):399-401

[48] Pereira AAJ, Sampaio-Neto R. A random-list based LAS algorithm for near-optimal detection in large-scale uplink multiuser MIMO systems. In: 19th Int. ITG Workshop Smart Antennas. Ilmenau, Germany: VDE; 2016

[49] Chaudhary M, Meena NK, Kshetrimayum RS. Local search based near optimal low complexity detection for large MIMO system. IEEE International Conference on Advanced Networks and Telecommunications Systems. Bangalore, India: IEEE; 2016

[50] Aravindan I, Kumar A, Snehith TC, Padmakumar A, Ramanathan R. A performance study of MIMO detectors in the presence of channel estimation errors. In: Int. Conf. Commu, Infor. And Computing Tech. (ICCICT). Mumbai, India: IEEE; 2015

[51] Chihaoui I, Ammari ML, Fortier P. Improved LAS detector for MIMO systems with imperfect channel state information. IET Communication. 2019;**13**(9):1297-1303

[52] Chihaoui I, Ammari ML. LAS detector with soft-output MMSE initialization under imperfect channel estimation and channel correlation. Wireless Personal Communication. 2019;**108**:213-220

[53] Mohammed SK, Zaki A, Chockalingam A, Rajan BS. High-rate space–time coded large-MIMO systems: Low-complexity detection and channel estimation. IEEE Journal of Selected Topics in Signal Processing. 2009;**3**(6): 958-974

[54] Qin Z, Xu J, Tao X, Zhou X. Improved depth-first-search sphere decoding based on LAS for MIMO-OFDM systems. In: IEEE 82nd Veh. Tech. Conf. (VTC2015-Fall). Boston, MA, USA: IEEE; 2015

[55] Chihaoui I, Ammari ML. Suited architecture for massive MIMO detector based on antenna selection and LAS algorithm. In: Int. Symp. Signal, Image, Video and Commu. (ISIVC). Tunis, Tunisia: IEEE; 2016

[56] Cerato B, Viterbo E. Hardware implementation of a low-complexity detector for large MIMO. In: IEEE Int. Symp. Circuits and Systems. Taipei, Taiwan: IEEE; 2009

[57] Li L, Hou H, Meng W. Convolutional-neural-network-based detection algorithm for uplink multiuser massive MIMO systems. IEEE Access. 2020;**8**:64250-64265

[58] Ullah A, Choi W, Berhane TM, Sambo Y, Imran MA. Soft-output deep LAS detection for coded MIMO systems: A learning-aided LLR approximation. IEEE Transactions on Vehicular Technology. 2024;**73**(9):13178-13192

[59] Wu Y, Gao X, Zhou S, Yang W, Polyanskiy Y, Caire G. Massive access for future wireless communication systems. IEEE Wireless Communication. 2020;**27**(4):148-156

[60] Wei Z, Liu F, Masouros C, Su N, Petropulu AP. Toward multi-functional 6G wireless networks: Integrating sensing, communication, and security. IEEE Communications Magazine. 2022; **60**(4):65-71

[61] Nelson LB, Poor HV. Iterative multiuser receivers for CDMA channels: An EM-based approach. IEEE Transactions on Communications. 1996; **44**(12):1700-1710

[62] Varanasi MK, Aazhang B. Near-optimum detection in synchronous code-

division multiple access systems. IEEE Transactions on Communications. 1991; **39**:725-736

[63] Wu B, Wang Q. New suboptimal multiuser detectors for synchronous CDMA systems. IEEE Transactions on Communications. 1996;**44**(7): 782-785

[64] Raphaeli D. Suboptimal maximum-likelihood multiuser detection of synchronous CDMA on frequency-selective multipath channels. IEEE Transactions on Communications. 2000; **48**(5):875-885

[65] Tanaka T. A statistical-mechanics approach to large-system analysis of CDMA multiuser detectors. IEEE Transactions on Information Theory. 2002;**48**:2888-2910

[66] Guo D, Verdú S. Randomly spread CDMA: Asymptotics via statistical physics. IEEE Transactions on Information Theory. 2005;**51**(6): 1983-2010

[67] Tse DNC, Hanly SV. Linear multiuser receivers: Effective interference, effective bandwidth and user capacity. IEEE Transactions on Information Theory. 1999;**45**:641-657

Chapter 6

Evaluating MIMO and Massive MIMO Performance with Rayleigh, Rician, and Nakagami Fading Channels along with Comparing Half-Duplex and Full-Duplex Modes Using HMR Protocol

Daphney Joann and Vayanaperumal Rajamani

Abstract

Multiple-Input Multiple-Output (MIMO) technology enhances wireless communication by utilizing multiple antennas, thus simplifying resource allocation through Time Division Multiplexing (TDM). MIMO effectively overcomes bandwidth limitations with channel state information (CSI) and multiple antennas. Massive MIMO (mMIMO), which employs large-scale antenna systems, further improves channel capacity, throughput, and energy efficiency. The performance of MIMO and Massive MIMO is evaluated across various channels. A new protocol, Heterogeneous Multiplex Relay (HMR), is proposed to enhance spectrum efficiency (SE) in MIMO systems using both Half Duplex (HD) and Full Duplex (FD) modes. The HMR protocol focuses on selecting relays between HD and FD to combat fading, thereby achieving both diversity and multiplexing gains. In Massive MIMO, the protocol incorporates Opportunistic Relay Selection, requiring minimal feedback for relay and mode selection. The max-min criteria identify the best channel, facilitating optimal relay selection. Additionally, Coded Cooperation and Successive Interference Cancelation (SIC) are considered to mitigate the loss of multiplexing gain. Simulations demonstrate that the HMR protocol achieves 80% capacity performance by efficiently allocating pathways connecting HD and FD. Compared to Rayleigh, Rician, and Nakagami Fading Channels, it also exhibits better Bit Error Rate (BER) versus Signal- to-Noise Ratio (SNR) performance.

Keywords: Multiple Input Mutiple Output (MIMO), Half Duplex (HD), Full Duplex (FD), Heterogenous Multiplex Relay protocol (HMR), Bit Error Ratio (BER), Signal to Noise Ratio (SNR)

1. Introduction

MIMO (Multiple-Input and Multiple-Output) technology in wireless communications utilizes multiple antennas at both the transmitter and receiver ends. The use of more antennas directly increases the channel's throughput and effectively provides the required bandwidth. MIMO enhances reliability by sending data through multiple paths, known as spatial diversity, and improves data rates by transmitting data over different propagation paths, termed spatial multiplexing. MIMO technology boosts the processing power, data rate, and reliability of networks. SISO (Single Input Single Output), MISO (Multiple Input Single Output), MOSO (Multiple Output Single Output), and MIMO were the several MIMO configurations.

The different MIMO configurations include Single Input Single Output, Multiple Input Single Output, Multiple Output Single Output, and Multiple Input Multiple Output. Massive MIMO, an extension of MIMO with a large number of antennas at both ends, achieves better throughput and spectrum efficiency. By increasing the number of antenna links, Massive MIMO significantly enhances network capacity and coverage. Transitioning from MIMO to Massive MIMO, particularly under Time Division Duplexing (TDD) operation, brings substantial improvements in throughput and energy efficiency due to the increased number of antennas.

Multiple-Input Multiple-Output (MIMO) technology enhances channel capacity and leverages multiple antennas to improve both multiplexing and diversity gains. MIMO effectively mitigates multipath fading through space diversity, as the multiple antennas capture different observations of the same signal. This work introduces a new Heterogeneous Multiplex Relay (HMR) Protocol aimed at optimizing the Diversity-Multiplexing Tradeoff (DMT) in a network that employs both half-duplex and full-duplex relaying. The HMR protocol's hybrid approach distinguishes between half-duplex and full-duplex modes based on Shannon Law Capacity, expressed as $C=B\log2 (1 + SNR)$. The protocol selects the appropriate mode by evaluating capacity and self-interference (SI). The performance of the HMR protocol is evaluated against different channel models, such as Rayleigh, Rician, and Nakagami, to demonstrate its effectiveness in diverse communication environments.

Multiple-Input Multiple-Output (MIMO) technology has advanced third and fourth generation wireless systems, addressing the increasing demand for high data rates, spectral efficiency, and reliability. This can be achieved using various techniques, including blind channel estimation, which allow for the estimation of channel parameters without requiring prior knowledge of the transmitted signals [1]. In multi-antenna wireless communications, spatial modulation is used to enhance signal processing at the base station (BS) receiver, with each user having antenna configurations, such as 2 x 2, 4 x 4, or 8 x 8. This approach is part of MIMO (Multiple-Input Multiple-Output) technology and can scale up to Massive MIMO systems with configurations like 64x64 or 128x128 antennas [2].

Massive MIMO, which uses a large number of antennas, aims to improve throughput and energy efficiency. At the base station, it involves using many parallel low-cost, low-power units that work effectively together, replacing the need for expensive hardware [3]. The performance of centralized and distributed deployments is analyzed both with and without cooperation. Using MIMO and Massive MIMO increases spectrum efficiency. A simple transmission scheme enhances user fairness by utilizing numerous base station antennas as data streams, which boost capacity [4]. A virtual antenna array with 64 elements operating at 3.33 GHz is used. Channel parameters, including the power delay profile and power delay spectrum, are analyzed based on measured data [5].

In wireless communications, additional technologies are required to address challenges like multipath propagation fading and Doppler shifts. To reduce the complexity of MIMO systems and lower the Bit Error Rate (BER), various detection algorithms are employed, including zero forcing (ZF), Minimum Mean Square Error (MMSE), and Maximum Likelihood (ML) [6]. Modulation techniques are used to transmit signals at higher frequencies, allowing multiple signals to share a single channel. The performance of 64 and 256 Quadrature Amplitude Modulation (QAM) is analyzed in terms of Bit Error Rate (BER), considering input signal, phase noise, and frequency offset. One prevalent kind of noise in the channel is Additive White Gaussian Noise (AWGN), which is a zero-mean Gaussian random process [7].

Rayleigh and Nakagami-m fading channels with Singular Value Decomposition (SVD) and the water filling technique are used to improve the spectral efficiency of MIMO systems [8]. A deterministic line-of-sight (LOS) path is part of this channel model, which employs Least-Square (LS), Element-wise MMSE (EW-MMSE), and Minimum Mean Square Error (MMSE) channel estimation methods [9].

Rayleigh fading and Additive White Gaussian Noise (AWGN) channels are used to examine the Bit Error Rate (BER) performance of Orthogonal Frequency Division Multiplexing (OFDM) systems with different Quadrature Amplitude Modulation (QAM) algorithms [10]. The BER performance of these modulation schemes is compared across AWGN, Rayleigh, and Rician channels [11]. Various spectral-efficient protocols have been proposed for half-duplex mode, including Amplify-and-Forward and Decode-and-Forward relay schemes. These protocols have been extended to support multiple terminals using the Orthogonalize-and-Forward (OF) relay method with a distributed zero forcing algorithm [12]. Additionally, a Full Interference Cancelation (FIC) algorithm has been developed to eliminate interrelay interference in two-trial cooperative systems [13].

Amplify-and-Forward relaying, combined with multi-antenna configurations, introduces a Random Sequential Relaying Scheme that achieves an enhanced Diversity-Multiplexing Tradeoff (DMT) [14]. In Full Duplex (FD) mode, interference arises during antenna selection, necessitating self-interference cancelation techniques in full-duplex heterogeneous networks to effectively double spectrum efficiency [15].

An Opportunistic Decode-and-Forward Relay Assortment System has been investigated in the context of free and similarly distributed Rayleigh and Nakagami fading channels [16]. This study introduces a Multiple Relay Coded Cooperation (MRCC) strategy that utilizes Decode-and-Forward relaying to mitigate signal loss in Half-Duplex communication scenarios [17]. Additionally, a hybrid Half-Duplex/Full-Duplex (HD/FD) communication approach in a relay-assisted network leverages throughput advantages and dynamically manages the transition between Half-Duplex and Full-Duplex based on the signal-to-noise interference ratio [18].

A Genetic Algorithm (GA)-based Transmit Antenna Selection (TAS) scheme is proposed to enhance self-backhaul capabilities, known as Integrated Access and Backhaul (IAB) [19]. Additionally, a Full-Duplex communication system that mitigates self-interference by focusing on line-of-sight (LOS) paths and utilizing multiple antennas achieves multiplexing gains [20].

The challenge of maximizing the minimum spectral efficiency (SE) is addressed by distributing the best possible uplink transmission power in a Massive MIMO (mMIMO) system that is cell-free (CF), user-centric (UC), and assisted by Rate Splitting Multiple Access (RSMA) [21]. With the increase in the number of antennas in Massive MIMO systems, traditional detection algorithms require enhancements. In order to overcome the performance constraints of conventional linear detectors and circumvent the high-dimensional matrix inversion of the linear Minimum Mean Square Error (MMSE) detector, the Broyden Quasi-Newton approach adds trainable variables, resulting in Broyden-Net [22].

This work presents several key contributions: it discusses the conversion of MIMO to Massive MIMO, with an emphasis on estimating fading channels in Rayleigh, Rician, and Nakagami environments. Additionally, in order to enable both half-duplex and full-duplex relaying networks, a heterogeneous multiplex protocol is developed. A dynamic switch between half-duplex and full-duplex modes is made by the collaborative relay selection according to Shannon Law Capacity. Full-Duplex is preferred when the capacity surpasses a predetermined threshold, and Half-Duplex is selected when the capacity falls below it. Furthermore, Half-Duplex is preferred in scenarios where self-interference occurs due to Full-Duplex operation.

This work is structured as follows. Section 2 introduces the System Model for MIMO and Massive MIMO. Section 3 covers the Protocol description, including Shannon Law Capacity, information on half-duplex and full-duplex operations, as well as the relay system based on the hybrid multiplex protocol. A comparison of the HMR protocol in Rayleigh, Rician, and Nakagami Fading Channels is given in Section 4. The findings and discussions are shown in Section 5. The Conclusion ends Section 6.

2. System model

Signals in wireless communication can fade due to factors like time and frequency dispersion, path loss, co-channel interference, and limited bandwidth. MIMO technology enhances spectral efficiency and improves transmission reliability. This work recommends using MIMO for multipath propagation and spectrum efficiency.

2.1 MIMO system model

In Multiple-Input Multiple-Output (MIMO) systems, a transmitter (Tx) sends data through multiple antennas to a receiver (Rx). The communication between the

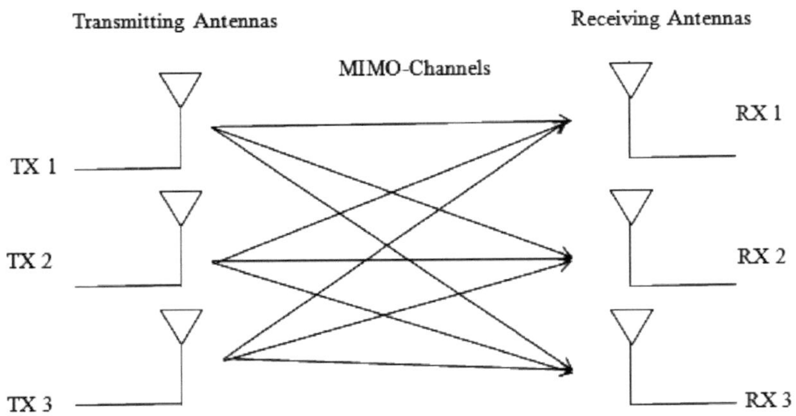

Figure 1.
MIMO System Model.

transmitter and receiver can be represented by a matrix that consists of N number of paths. This matrix captures the relationship between the multiple transmitting and receiving antennas, with each element of the matrix representing the path or channel between a particular pair of transmitter and receiver antennas. **Figure 1** typically illustrates this concept by showing the multiple antennas at both the transmitter and receiver ends and the various paths connecting them.

The MIMO System is modeled as.

$$y = Hx + n \tag{1}$$

where y represents the vector of received signals. x represents the vector of transmitted signals. H is the channel matrix that characterizes the effect of the transmission path between the transmitter and receiver antennas. n is the noise vector that accounts for the noise introduced in the communication system.

2.2 Massive MIMO system model

In Massive MIMO systems, there are hundreds of antennas at the base station, which simultaneously serve multiple users over the same frequency. Time Division Duplexing (TDD) operation is preferred due to the large number of antennas. Because of the extensive number of base station antennas, linear processing techniques become nearly optimal. There are two key schemes for linear processing in Massive MIMO: Linear Precoding in Uplink: This involves processing the signals received from multiple users to efficiently combine them and reduce interference. Linear Precoding in Downlink: This involves processing the signals before transmission to multiple users to ensure they are transmitted in a way that minimizes interference and optimizes signal quality.

2.3 Steps of uplink operation

The base station in a Massive MIMO system estimates the channel conditions based on requests from users. It uses pilot sequences to encode the data for transmission. By utilizing the estimated Channel State Information (CSI), the base station can efficiently decode and detect the data streams during communication.

2.4 Steps of downlink operation

In the downlink of a communication system, two important techniques are used: Beamforming that directs data streams to specific locations, allowing them to use the same frequencies at the same time, and Precoding that involves sending multiple data streams to increase overall data throughput.

3. Protocol description

The protocol operates using the MIMO Decode-and-Forward (DF) relay networks, following the max-min criterion to identify the optimal channel and select the best relay. This approach minimizes the need for extensive feedback and signaling to choose the relay and operation mode. The HMR protocol functions in two modes: Half Duplex Mode and Full Duplex Mode within a two-way relaying scheme. To switch between these modes, the protocol compares their reliability and capacity, using Shannon Law Capacity as a benchmark.

When switching to Full Duplex Mode, the wireless system's coded cooperation faces self-interference due to simultaneous signal transmissions. This interference can be managed by reducing the strength of the received signals, allowing for concurrent transmission. The HMR protocol helps improve the outage probability, resulting in higher data rates and better multiplexing gain.

3.1 Relays in duplex section

Signals can be received and retransmitted between the sender and the recipient with the help of relays. The HMR relay, based on the Decode-and-Forward (DF) method, handles encoding and decoding of signals. It is equipped with two antennas: a transmitter antenna and a receiver antenna. These basic relay operations are illustrated in **Figure 2**.

In the initial step, the source terminal S sends data via signal X to both the relay and the destination terminal, through either the channels h_{SR}, h_{SD}. The signals received at the relay and destination terminal can be represented as follows:

$$Y_{SR} = h_{SR}X + Z_{SR} \tag{2}$$

$$Y_{SD} = h_{SD}X + Z_{SD} \tag{3}$$

where Y_{SR} and Y_{SD} represent the communication between the source and the relay, and between the source and the destination, in that order. The relay terminal decodes the received signal Y_{SR} using Eqs. (2) and (3), re-encodes it, and then sends it to the destination (Y_{SD}). The representation of the signal from the relay to the destination Y_{RD} is given as follows:

$$Y_{RD} = gh_{RD}Y_{SR} + Z_{RD} \tag{4}$$

where Z_{RD} is the independent Gaussian noise related to the channel gains from the relay to the destination and gh_{RD} is the channel gain between the relay and the destination.

3.2 Shannon law capacity

Shannon Law Capacity calculates the channel capacity based on the average received signal power and bandwidth. This calculation helps determine whether to

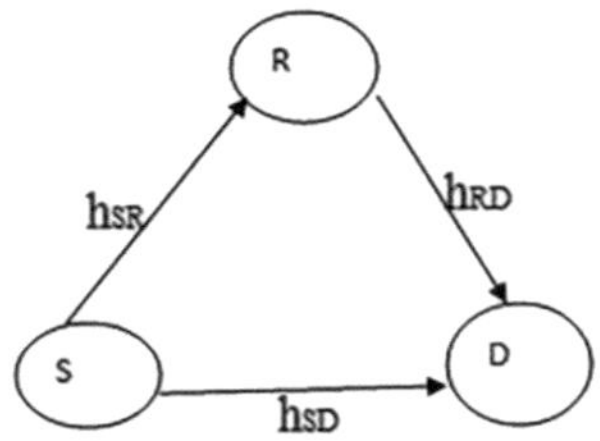

Figure 2.
Relays in duplex system.

switch between half-duplex and full-duplex relaying schemes. It is represented as follows:

$$C = Blog_2(1 + SNR) \tag{5}$$

In the context of the HMR protocol:

- The channel capacity C is determined by the available bandwidth B relative to the noise and interference levels.
- If $C > R$ where R represents a threshold related to system capabilities, the two-way relaying scheme operates in Full Duplex Mode.
- Conversely, if $C < R$, the relaying scheme operates in Half Duplex Mode.
- The HMR protocol achieves optimal SNR performance typically ranging from 15 to 40 dB, enhancing spectrum efficiency by effectively mitigating interference through techniques that subtract interfering signals from received signals, thereby minimizing self-interference.

3.3 Heterogeneous multiplex relay protocol

The two-way relaying scheme of the HMR protocol involves the exchange of information between User 1 and User 2 using Multiple-Input Multiple-Output (MIMO) systems, facilitated by multiple relays. Specifically, User 1 and User 2 exchange their information through a set of relays denoted as R_i where ($i = 1 \ldots N_R$). The sender and the recipient exchange two sets of messages, M1 and M2, in this system. Request and response messages are the terms used to describe the messages that are sent from the source to the relay and then from the relay to the destination. Channel coefficients hi, which have zero mean and unit variance and follow a Gaussian distribution, define the communication linkages between the nodes. This implies

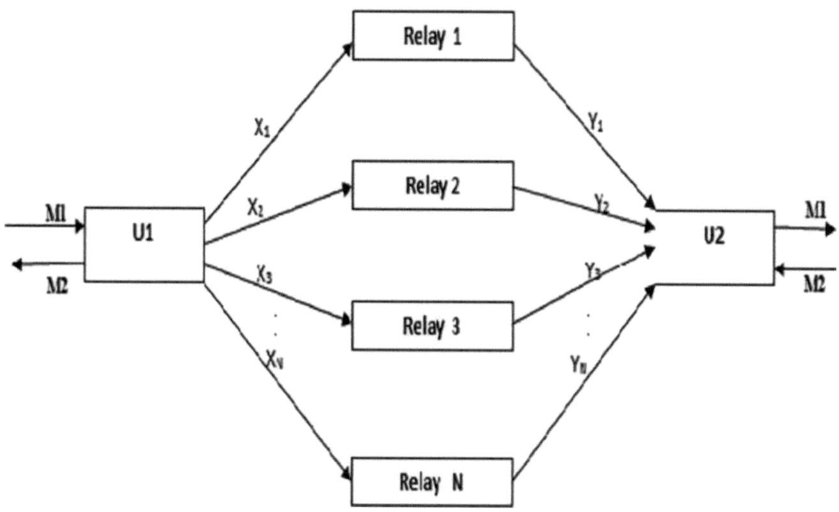

Figure 3.
Half-duplex and full-duplex relay systems.

that the channel conditions are modeled as random variables with a normal distribution, providing a realistic representation of the wireless communication environment. The HMR protocol employs a decode-and-forward mechanism. This process involves: encoding the message at the source, decoding the message at the relay, re-encoding the message at the relay, and finally, decoding the message at the destination (**Figure 3**).

3.3.1 Selection of the best relay

The transmission is performed in frames consisting of Z symbols. Each frame represents a block of symbols that are transmitted during a specific time interval.

$$Y_{R_j} = \sqrt{PS_i}\, hZ_iX_i + K_{X_i} \tag{6}$$

The decode-and-forward protocol uses the encoding and decoding procedure for transmission and reception of data using the coding symbols KX_i. The totally received signal in the destination is given as follows: YR_j represents the signal received in the relay, PS_i indicates the power scaling of the duplex, and X_i represents the user that directs the message from source to destination. The channel coefficients between source to relays and relays to destinations are related to the channel gain h and Gaussian Noise Vector Z_i.

$$Y_{D_i} = \sqrt{PR_j}\;\, hZ_i\, f\left(Y_{R_j}^{best}\right) + K_{Y_i} \tag{7}$$

where Y_{Di} symbolizes the overall signal received at the destination. Where PR_j denotes the power constraint concerning relay. The $f\left(Y_{R_j}^{best}\right)$ designates the function applied to find the best relay among Y_i. The K_{Yi} symbol represents the encoding and decoding functions with respect to Y_i.

3.3.2 Half duplex relay system

In the two-way Half Duplex relaying two time slots is required for data forwarding from source to destination because it does not support data transfer at the same time. The communication from source to relay is represented by the first time slot. The data transmission from relay to destination is represented by the second. The initial time slot is

$$Y_{SR_I} = \sqrt{PS_i}\, hZ_iX_i + K_{r_i} \tag{8}$$

The second time slot is

$$Y_{RD_i} = \sqrt{PS_i}\, hZ_iY_i + K_{Y_i} \tag{9}$$

where Y_{SRi} and Y_{RDi} denote the time slots for the relay to destination and the source to relay. The HMR protocol's operational flowchart is shown in **Figure 4**. It represents the initiating process between the Half Duplex and Full Duplex.

Algorithm for Heterogeneous Multiplex Relay Protocol.

Step 1: Define the set of users in the MIMO system as $U_i = [U_1, U_2 \ldots \ldots \ldots U_n]$.

Step 2: Specify $X_i = [X_1, X_2 \ldots . X_n]$ as the path from each relay to the destination.

Step 3: Define $Y_i = [Y_1, Y_2 \ldots . Y_n]$ by the path from relay to destination.

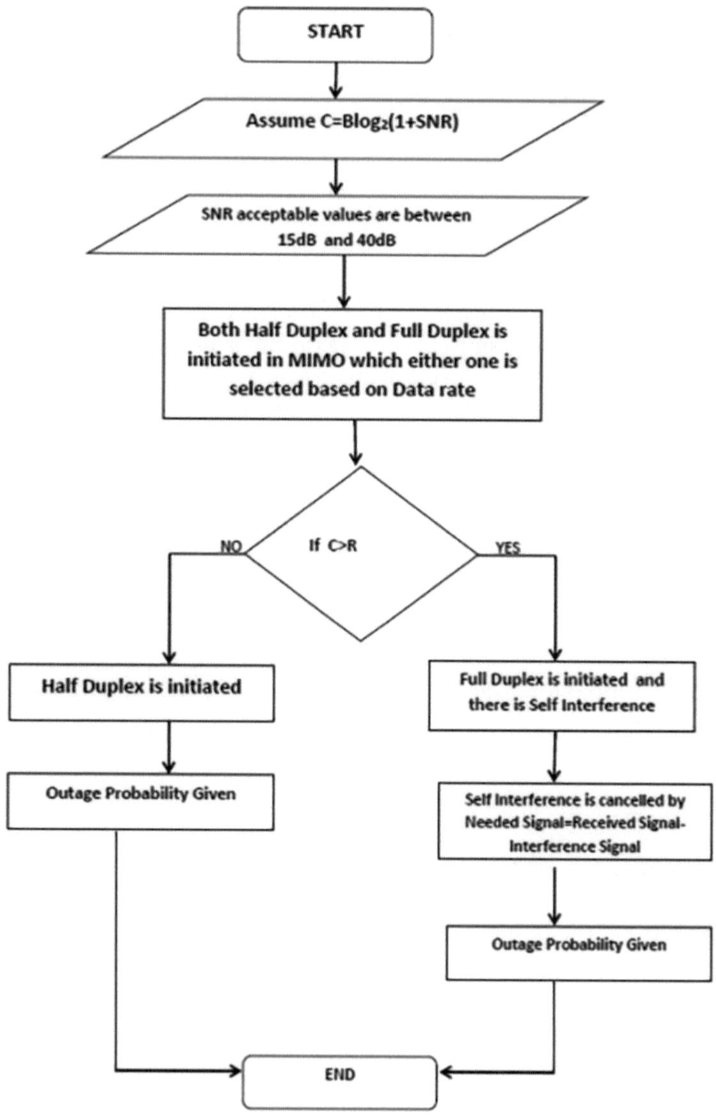

Figure 4.
An operational flowchart of the HMR protocol.

Step 4: Let $f\left(Y_{R_j}^{best}\right)$ be the selecting relay from $[Y_1, Y_2 \ldots . Y_n]$.

Step 5: Define channel capacity C as $C = B\log_2(1 + SNR)$ where.

B is the bandwidth and SNR is the signal-to-noise ratio.

Step 6: There are two operational modes for the channel capacity calculation: Half Duplex and Full Duplex.

Step 7: If $C > R$.

Step 8: Choose "Full Duplex" mode.

Step 9: Else.

Step 10: If $C < R$.

Step 11: Then select "Half Duplex" mode.

Step 12: In Full Duplex mode, enable "Self Interference."

Step 13: Use Self-Interference Cancelation to enhance the channel capacity.

Step 14: Calculate the Desired Signal by subtracting the Interference Signal from the Received Signal.

Step 15: End the process.

3.3.3 Full duplex relay system

Only full-duplex data transmission is needed in a two-way Full Duplex relay for a single time slot. Data can flow simultaneously on the same frequency. Considerable spectrum competency can be achieved using two-way FD relaying, but processing complexity is considerable as well. Self-interference resulting from co-channel conveyance is the cause of this effect. Self-interference cancelation works by using the whole received signal to cancel out the interference signal. The relay's received signal is

$$Y_{D_i} = \sqrt{PS_i}hZ_iX_i + \sqrt{PS_i}hZ_{i-1}Y_i + f\left(YR_j^{best}\right) + K_{r_i} + K_{Y_i} \tag{10}$$

Y_{Di} represents the signal of full duplex received in the time slot, X_i and Y_i are the users between the sender and the receiver. $f\left(Y_{Rj}^{best}\right)$ designates the function applied for finding the best relay among Y_i. The K_{Yi} and K_{ri} symbols represent the encoding and decoding functions.

4. Estimation of different channels

4.1 Rayleigh channel

When leading propagation does not occur in the line of sight between the transmitter and the receiver, Rayleigh fading can be used in wireless environments. It offers both ionospheric and tropospheric signal propagation. The channel, which is inversely proportional to the wavelength power, represents the scattering of electromagnetic radiation.

MIMO Rayleigh channel: A MIMO system's capacity across a Rayleigh fading channel is dependent on the number of broadcast and receiver antennas. The water filling procedure, which determines the singular value of the Rayleigh channel matrix, provides the basis for MIMO power transmission.

Channel Rayleigh using Massive MIMO: In order to arrange multiplex users based on time in the Rayleigh channel, massive MIMO, which is connected to Long-Term Evolution (LTE) advanced, is outfitted with a huge number of antennas M and Base Stations (BSs) with spatial resolution. The Zero forcing technique is used to reduce interuser interference.

4.2 Rician channel

A "rician channel" is a strong signal that appears when there is a line-of-sight signal. The Rician distribution is used to characterize the amplitude gain that is attained. According to the Rician model, a dominant wave is a deterministic process that is susceptible to shadow attenuation. This dominant wave may be the phasor sum of two or more dominating signals.

Rician Channel with MIMO: Perfect channel state information (CSI) is used to observe the Rician channel with multiple inputs and multiple outputs on both the transmitter and receiver sides. The precise statement of Ricardian fading, which indicates when the fading coefficients are independent but not necessarily distributed, is then provided by the CSI.

Rician Channel with Massive MIMO: Rician fading Massive MIMO using spatially correlated channels. A deterministic line-of-sight (LOS) path defining a spatially correlated multipath environment makes up the channel model.

4.3 Nakagami channel

Signals that are not necessary in line of sight are received by the Nakagami channel model using both diffuse and specular scattering. Less amplitude is contributed by the gamma or fading distribution. With parameter m shape factor, the Nakagami exhibits distinct behavior. The Nakagami distribution exhibits Rayleigh-like behavior, when $m = 1$. The Nakagami functions as a Rician fading channel, when $m = 0$. In contrast to Rayleigh fading, the signal strength decreases, when $m > 1$.

Nakagami Channel with MIMO: A multiuser scheduler that boosts system performance by increasing the number of antennas is the Nakagami channel with MIMO systems. For a number of multiuser scheduling methods, the equations with the probability density function (PDF) and cumulative distribution function (CDF) are derived in order to examine the performance.

Nakagami Channel with Massive MIMO: Massive MIMO usually refers to many antenna arrays that serve many single antennas. The performance is measured based on the parameter m with two schemes of linear detection, namely maximum ratio combining (MRC) and zero forcing receiver (ZF). As m increases, the spectral efficiency increases slowly with the rise in the number of base station antennas. The ZF shows better performance.

5. Results and discussions

Based on SNR phases, the numerical results of Bit Error Rate (BER) and Eb/NO are compared. The self-interference that happens in Full Duplex relaying systems is thought to be the cause of the variances in performance across Half- and Full Duplex relaying systems. Considered is a radio that operates at 2.5 GHz with 64 QAM. An average of the Signal-to-Noise Ratio between 0 and 30 dB is compared. When the signal strength is between 20 and 30 dB, it is good. Bit Error Rate averages that are acceptable are going to range between 10^{-1} and 10^{-9}. The starting values for the number of antennas will be given as follows: 2 antennas with 2 users reach 3 SNR with a capacity of 2 bits/S/Hz, 4 antennas reach 10 dB with a capacity of 2 bits/S/Hz, 8 antennas reach 15 dB with a capacity of 5.5 bits/S/Hz, and 16 and 32 antennas reach 20 to 30 dB with a capacity of 5.5 to 8 bits/S/Hz. Transmission in parallel enables it feasible to accomplish Full Duplex performance, which is superior to Half Duplex performance. Self-interference (SI) causes the performance to deteriorate, while successive cancelation/suppression causes it to improve. Typically, the Rayleigh, Rician, and Nakagami-m fading channels are used to construct the BER analysis.

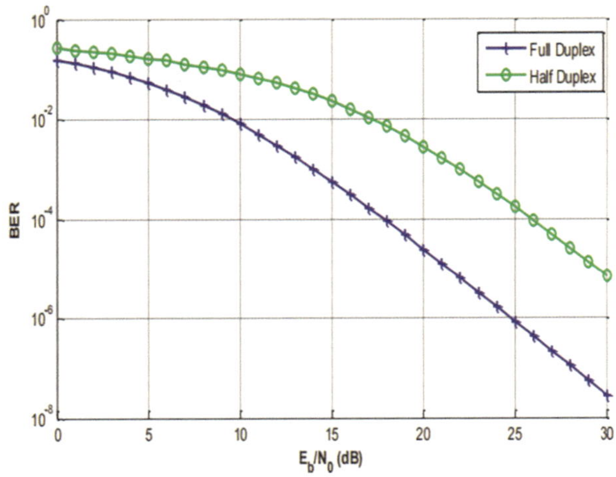

Figure 5.
BER vs. EB/No performance of Half Duplex and Full Duplex in HMR protocol.

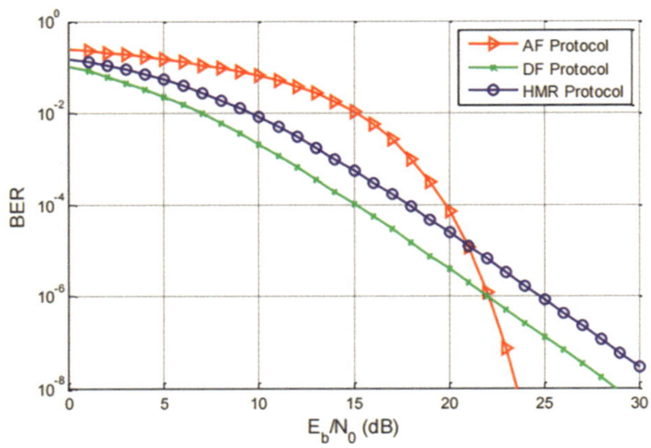

Figure 6.
BER vs. SNR comparison of Half Duplex in AF, DF, and HMR protocol.

The HMR protocol describes the hybrid behavior that shifts between Half Duplex and Full Duplex Mode based on Shannon Law Capacity, and the Time Division Multiple Access (TDMA) technique delivers a good performance. **Figure 5** displays the BER vs. EB/No performance. The findings illustrate the performance of the Half Duplex and Full Duplex in the HMR protocol, with an average of 0 to 30 dB and BER ranging from 10^0 to 10^{-5} for Half Duplex and 10^0 to 10^{-7} for Full Duplex.

The Half Duplex BER vs. SNR comparison in AF, DF, and HMR protocol is displayed in **Figure 6**. Faster data speeds are offered by 64 QAM, which raises the Bit Error Rate from 10^0 to 10^{-8}. From 0 to 30 dB, the SNR number shows good signal strength. With a Bit Error Rate of 10^{-8}, the Amplify-and-Forward AF Protocol in Half Duplex with 64 QAMs achieves an SNR value of 24 dB, offering sufficient signal strengths. A BER analysis of the Decode-and-Forward DF protocol in Half-Duplex relays using MIMO asymmetric relays shows that the SNR value approaches 28 dB. A superior outage comparison between AF and DF protocols delivering good signal strength and tolerating SNR values of 30 dB is provided by the currently proposed protocol, HMR.

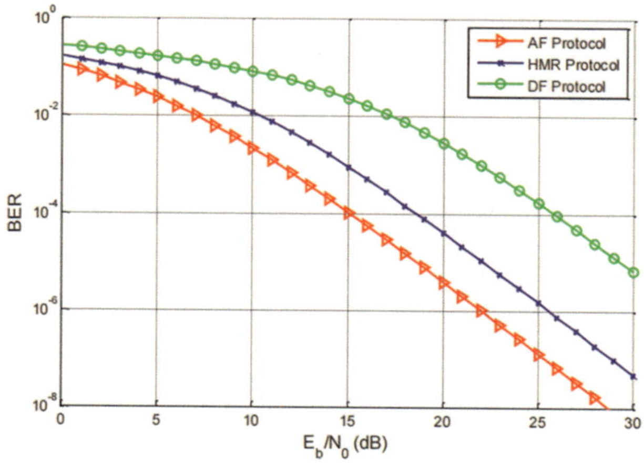

Figure 7.
BER vs. SNR comparison of Full Duplex in AF, DF, and proposed HMR protocol.

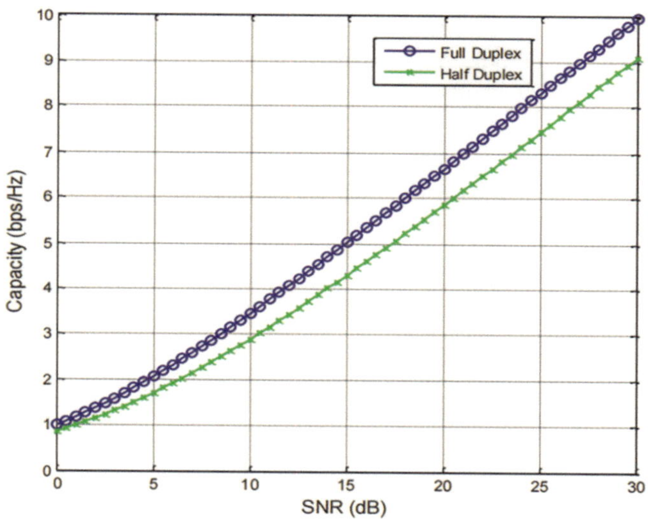

Figure 8.
Ergodic Capacity of Half Duplex and Full Duplex in the HMR protocol.

The Full Duplex BER vs. SNR comparison in the AF, DF, and proposed HMR protocol is displayed in **Figure 7**. The transfer of data is possible up to an SNR value of 28 dB using the Full Duplex with Amplify-and-Forward protocol. Sixty-four QAM is the one used in this instance. SNR levels in the Forward and Decoder protocols exceed up to 30 dB. Concurrent transmission is possible using the HMR protocol's Full Duplex, coded collaboration, and sequential interference cancelation. BER transmission is allowed up to 30 dB SNR values, allowing 64 QAM to achieve higher data speeds up to 10^{-7}.

Ergodic capacity contrasts the performance of half-duplex and full-duplex relaying systems and examines channel capacity. The primary goal of capacity analysis is to give wireless communication performance metrics that maximize throughput. The Ergodic Capacity of Half Duplex and Full Duplex in the HMR Protocol is displayed in **Figure 8**. In this case, the half-duplex outage capacity is 8 to 9 bps/Hz, while the full-duplex outage capacity is 9 to 10 bps/Hz. Performance wise, the Full Duplex outperforms the Half Duplex, as indicated in Hertz.

6. Conclusion

In this work, we introduced a novel maximum spectrum efficiency Heterogeneous Multiplex Relay Protocol in MIMO. Our proposal in this work is a new Heterogeneous Multiplex Relay Protocol in MIMO that maximizes the efficiency of the spectrum between Half Duplex and Full Duplex. The suggested protocol allots the time appropriately to guarantee no relay loss while preserving half-duplex and full-duplex functionality with time slots. Throughput enhancement and significant multiplexing gain are demonstrated by the DMT analysis of the HMR protocol with different antenna situations. Rayleigh, Rician, and Nakagami channels' performances are evaluated in terms of BER vs. SNR, and Nakagami performs better between Half Duplex and Full Duplex. By comparing BER vs. SNR and Ergodic capacity between Half Duplex and Full Duplex, simulation findings demonstrate HMR performance. HMR protocol's allocated channels between Half Duplex and Full Duplex allow for 80% capacity performance. Every fading channel has unique properties, and achieving SNR values aids Nakagami and Rayleigh in demonstrating a respectable diversity gain. As a result, the HMR protocol offers a strong relay network with maximum capacity.

Author details

Daphney Joann[1*] and Vayanaperumal Rajamani[2]

1 Department of Computer Science and Engineering (Data Science), Sri Venkateswara College of Engineering and Technology, Chittoor, India

2 Department of Electronics and Communication Engineering, Veltech Multitech Dr. Rangarajan Dr. Sakunthala Engineering College, Chennai, India

*Address all correspondence to: daphneyjoann@gmail.com

References

[1] Pappa M, Ramesh C, Madhushri NK. International Conference on Wireless Communications, Signal Processing and Networking (WiSPNET). 2017. pp.1808-1812. DOI: 10.1109/WISPNET.2017.8300073

[2] Lakshmi Narasimhan T, Raviteja P, Chockalingam A. Large-Scale Multiuser SM-MIMO Versus Massive MIMO. In: 2014 Information Theory and Applications Workshop (ITA) Conference. February 2014. pp.4-9

[3] Larsson EG, Edfors O, Tufvesson F, Marzetta TL. Massive MIMO for next generation wireless systems. IEEE Communications Magazine. 2015;**52**(2):186-195

[4] Dierks S, Zirwas W, Jager M, Panzner B, Kramer G. MIMO and massive MIMO —Analysis for a local area scenario. In: 23rd European Signal Processing Conference (EUSIPCO). 2015. pp. 2496-2500

[5] Fei D, He R, Ai B, Zhang B, Guan K, Zhong Z. Massive MIMO channel measurements and analysis at 3.33 GHz. In: 10th International Conference on Communications and Networking in China (China Com). 2015

[6] Azad A, Suhel Md., Jabeen U, Pathak A. Study of various channel estimation techniques in OFDM mobile wireless channel. International Journal of Emerging Technologies and Innovative Research (JETIR). 2015;**2**(2):271-278

[7] Sadinov SM. Simulation study of M-ARY QAM modulation techniques using Matlab/Simulink. In: 2017 40th International Convention on Information and Communication Technology. Electronics and Microelectronics (MIPRO). 2017. pp. 547-554

[8] Solanki M, Gupta S. Performance Analysis of Nakagami and Rayleigh Fading for 2 × 2 and 4 × 4 MIMO Channel with Spatial Multiplexing. In: Future Internet Technologies and Trends. First International Conference, ICFITT 2017, Surat, India. 2017. pp. 254-263

[9] Ozdogan O, Bjornson E, Larsson EG. Massive MIMO with spatially correlated Rician fading channels. IEEE Transactions on Communication. May 2019;**67**(5):1-17

[10] Farzamnia A, Hlaing NW, Mariappan M, Haldar MK. BER comparison of OFDM with M-QAM modulation scheme of AWGN and Rayleigh Fading channels. In: 2018 9th IEEE Control and System Graduate Research Colloquium (ICSGRC 2018). Shah Alam, Malaysia; 3–4 August 2018. pp. 54-58

[11] Usha SM, Nataraj KR. BER performance of digital modulation schemes with and without OFDM model for AWGN, Rayleigh and Rician channels. International Journal of Science and Research (IJSR). 2015;**4**(11):330-335

[12] Rankov B, Wittneben A. Spectral efficient protocols for half-duplex fading relay channels. IEEE Journal on Selected Areas in Communications. 2007;**25**(2):379-389

[13] Luo C, Gong Y, Zheng F. Full interference cancellation for two-path relay cooperative networks. IEEE Transactions on Vehicular Technology. 2011;**60**(1):343-347

[14] Gharan S, Bayesteh A, Khandani A. Diversity-multiplexing tradeoff in multi-antenna multi-relay networks:

Improvements and some optimality results. IEEE Transactions on Information Theory. 2013;**59**(6):3892-3914

[15] Hong S, Brand J, Choi J, Jain M, Mehlman J, Katti S, et al. Applications of Self-Interference Cancellation in 5G and Beyond. IEEE Communication Magazine. 2014;**52**(2):114-121

[16] Sarker M, Fapojuwo AO. Uplink power allocation for RSMA-aided user-centric cell-free massive MIMO systems. In: 2023 IEEE 97th Vehicular Technology Conference (VTC2023-Spring). 2023. pp. 1-4

[17] Hong B, Choi W. Overcoming half-duplex loss in multi-relay networks: multiple relay coded cooperation for optimal DMT. IEEE Transactions on Communications. 2015;**63**(1):66-78

[18] Chen H, Zhao F. A Hybrid half-duplex/full-duplex transmission scheme in relay-aided cellular networks. EURASIP Journal on Wireless Communications and Networking. 2017; **1**:1-10

[19] Akif F, Malik A, Qureshi I, Abassi A. Transmit and receive antenna selection based resource allocation for self-Backhaul 5G Massive MIMO HetNets. The International Arab Journal of Information Technology. 2021;**18**(6): 755-766

[20] Chae S, Lee K. Cooperative Relaying for Multi-User MIMO Wireless Backhaul Networks. IEEE Transactions on Vehicular Technology. 2021;**70**(3):2794-2806

[21] Yu Y, Zhang S, Ying J, Wang P. Massive MIMO detection method based on quasi-newton methods and deep learning. IEEE Communications Letters. 2024;**28**(4):912-916

[22] Zhang Z, Chai X, Long K, Vasilakos A, Hanzo L. Full-duplex techniques for 5G Networks: Self-interference cancellation, protocol design, and relay selection. IEEE Communications Magazine. 2015;**53**(5):128-137

Section 4

MIMO Antennas

Chapter 7

Massive MIMO Antenna Arrays for Low-Frequency Bands

Kerlos Atia Abdalmalak, Ahmed El Yousfi and Daniel Segovia Vargas

Abstract

5G and future mobile communications increasingly emphasize massive Multiple-Input Multiple-Output (MIMO) technology to meet the demanding requirements for higher capacity and data rates. Solutions for high-frequency bands above 1 GHz exist, but increasing MIMO order in low-frequency bands remains unexplored despite their advantages as wider coverage. This chapter summarizes the challenges of designing base station antennas for massive MIMO at 5G New-Radio (5G-NR) 700 MHz bands, transitioning from 4T4R to 16T16R dual-polarized MIMO without increasing overall size. Hence, compact antennas are required without sacrificing the array performance, especially in terms of broadband, gain, half-power beamwidth (HPBW), and isolation. The main requirements for a proper radiating element are summarized. Three candidates are presented based on ultrawideband and resonating structures (including spirals, differential patches, and Dielectric Resonator Antennas (DRAs)) followed by performance studies in isolated and subarray configurations. A differential metallic cap-loaded multi-layer DRA element proves to be a strong candidate. A truncated-version of three subarrays is investigated demonstrating outstanding capability with broadband, flat gain of 8 dBi, and average HPBW below 35°. Also, potential extension from 4T4R to 32T32R is analyzed, which remains unattainable with other solutions at low frequencies. Finally, the fabrication and measurement of a preliminary element are discussed.

Keywords: MIMO MIMO antennas, compact high-gain antennas, planar spiral, L-feed differential patch, dielectric resonator antenna (DRA)

1. Introduction

Future Multiple-Input and Multiple-Output (MIMO) systems aim to increase capacity by introducing more radiating elements to form massive MIMO arrays. These multiple numbers of antennas are able to narrower the steered beam to optimize the signal at the receiver terminal [1]. Several systems are proposed for higher frequencies, usually higher than 1GHz [2, 3]. However, still a big room for enhancement for massive MIMO at the lower band as an essential solution for achieving wider coverage areas and, hence, connecting a massive amount of devices (for example, for internet of things IoT services). The challenges of having massive MIMO for 5G and beyond base stations at low-frequency bands are devoted mainly to three terms. First, having a compact radiated element, hence, the array can include a larger number of transceivers (TRX) without increasing the total size of the overall system which is restricted to standard sizes. This is also mandatory for enhancing the isolation between the array elements. Secondly, obtaining a small HPBW means higher directivity is needed to be able to scan a wide range of angles. And finally, covering broadband to have higher data rates.

To achieve the aforementioned goals several works have been done in recent years. Conventional geometries for wideband dual-polarized antennas that are commonly used for 5G base station applications are proposed such as dipoles [4, 5], patches [6, 7], and slot antennas [8]. However, mostly, the size of these antennas is half wavelength or even exceeds it at the working frequency. Therefore, the need for developing new compact antennas [9] with broadband radiation is of great importance. Generally, miniaturization can be achieved by inserting meandering lines in patches [10], loading Complementary Split Ring Resonator (CSRR) monopole [11], or high permittivity dielectric resonator [12]. The latter technique is used in designing many dual-polarized compact antennas for massive MIMO 5G base station applications [13, 14]. Although the size is reduced, the main issue of such designs is their narrow bandwidth ($< 5\%$), and low gain. To widen the bandwidth with a compact size and a relatively high gain, a dual-polarized dielectric resonator magnetoelectric antenna loaded with a metallic loop [15] is proposed, which leads to an improved bandwidth of 14.5% and a gain of 7 dBi. Even though the antenna exhibits a compact size of $0.33\lambda \times 0.33\lambda$, there is still room for enhancing both the bandwidth and compactness of the radiating element. Another way to achieve broadband is based on loading metasurface with DRA to form a hybrid system in which two close resonant frequencies are excited, one associated with the DRA whereas the other one is due to the metasurface [16, 17]. As a result, a broad bandwidth of typically 17% [16], and 15.5% [17] is achieved, but at the expense of the large size of $0.5\lambda \times 0.5\lambda$, [16].

Most of the previous works have a large size which makes their integration in massive MIMO array scenarios (as shown in **Figure 1**) a tough task. This becomes very challenging when working with a low-frequency band. To implement four dual slant radiating elements in the vertical column of 500 mm to form a subarray, a compact antenna is needed at 700 MHz (As shown in **Figure 1**). Additionally, most of the

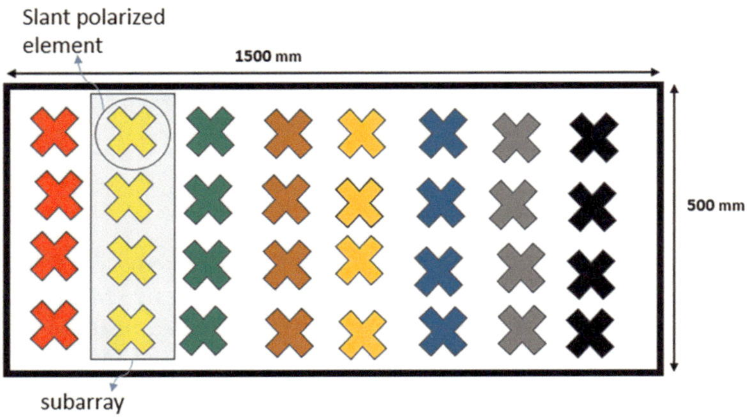

Figure 1.
Architecture of the proposed massive MIMO 5G base station array at low band.

simulations in this chapter are undertaken for only three subarrays due to the large computational time of the entire array and because in such a case, all the electromagnetic interactions between the neighbor's subarrays are taken into account, as well as the edge effects.

In this chapter, a summary of compact, high-gain broadband elements will be discussed. Then several potential solutions for the radiated element will be presented in addition to showing the results of both isolated performance and in an array configuration.

2. Compact broadband single radiating element

As a summary of the previous section, to implement massive MIMIO at low-frequency bands with the same current size restrictions of base stations, the main requirements for the antenna element are

- Bandwidth: covering broadband from 700 to 800 MHz.
- Polarization: dual slant linear polarization $(\pm 45^{\circ})$.
- Size: Compact element with a size smaller than $0.23\lambda_{fmin} \times 0.23\lambda_{fmin}$ which is corresponding to 100 mm × 100 mm for a minimum frequency of 700 MHz.
- Gain: high gain above 6 dBi to produce a smaller HPBW.
- Height: low profile smaller than $0.23\lambda_{fmin}$ which is corresponding to 100 mm.

After testing several radiation elements, the section provides the best-optimized elements that cover these requirements.

2.1 Planar spiral element

2.1.1 Single-polarization spiral

In this section, a self-complementary planar spiral antenna will be presented for our band of interest. This first study will involve a single circular polarization as an initial stage, which will be updated later in this section to consider the actual requirement of dual polarization. It is worth noting also that despite the base station MIMO systems requiring dual linear polarization $(\pm 45^{\circ})$, once we have dual circular polarization, polarization conversion can be performed using extra circuits in real time [18]. The spiral structure is drawn in **Figure 2**, to provide a compact design, the spiral arms are printed on a high permittivity substrate made of Arlon AR1000 with $\epsilon_r = 10$ and $\tan\delta = 0.003$.

The spiral arms structure ends are short-circuited, providing more miniaturization [19]. As shown in **Figure 3**, the spiral covers broadband even higher than the required one with an impedance-matching bandwidth from 650 MHz to 1.1 GHz and a flat gain of about 7 dBi. The simulations are performed using both the time-domain solver of Computer Simulation Technology (CST) [20] and High-frequency Simulation Software (HFSS) [21]. It is true that the axial ratio bandwidth (**Figure 4**) where we have a

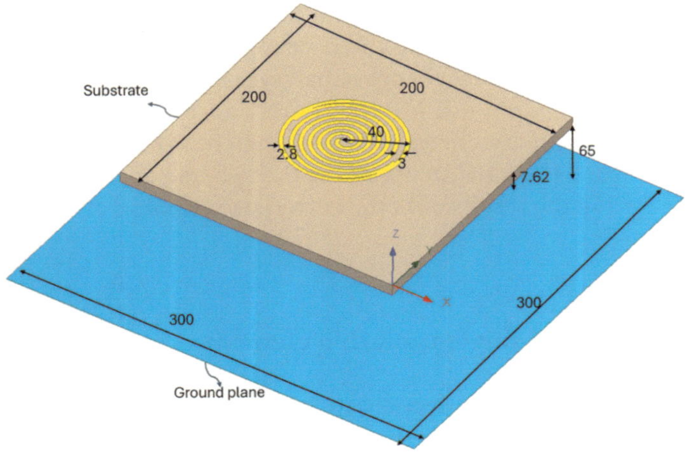

Figure 2.
Spiral radiating element structure with dimensions in mm.

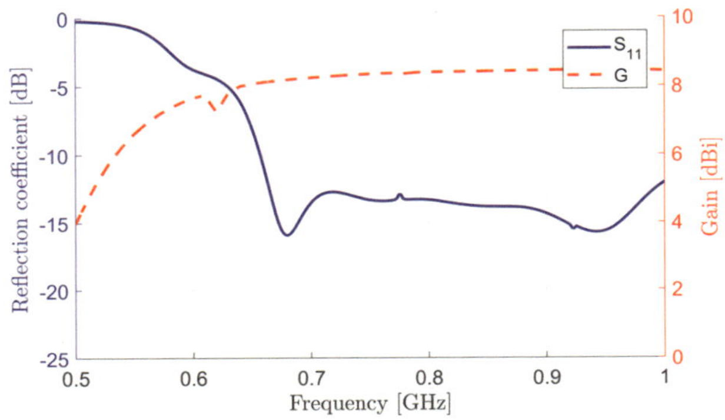

Figure 3.
S_{11} and gain of the spiral element.

circular polarization ($AR \leq 3$ dB) is slightly smaller; however, the impedance and axial ratio product covers the range from 700 to 920 MHz which is large enough for supporting MIMO low-frequency base stations. This good performance is obtained with a compact spiral diameter of 81 mm $\left(0.19\lambda_{fmin}\right)$ printed on 200 mm × 200 mm substrate.

By default, the planar spiral radiates bidirectionally, so this right-handed spiral radiates right-hand circular polarization (RHCP) in the upper parts while losing half of the power in radiating left-hand circular polarization (LHCP), and this strongly reduces the gain of about 3 dBi. Conical spiral can completely fix this limitation and provides a very high gain of about 10 dBi [22], however, the required height is significantly large (about $1.1\lambda_{fmin}$ [23]), which can not be afforded here for this low frequency due to the size limitations of the actual base stations. Consequently, a metallic reflector is used here to provide unidirectional radiation patterns with a good front-to-back ratio (FBR) of more than 23 dBi, as demonstrated in **Figure 5**. The reflector is introduced at a distance of 65 mm from the spiral and has a size of 300 mm × 300 mm. The large sizes of substrate and ground here are not an issue, as this will be shared for all array elements later once the array version is under consideration.

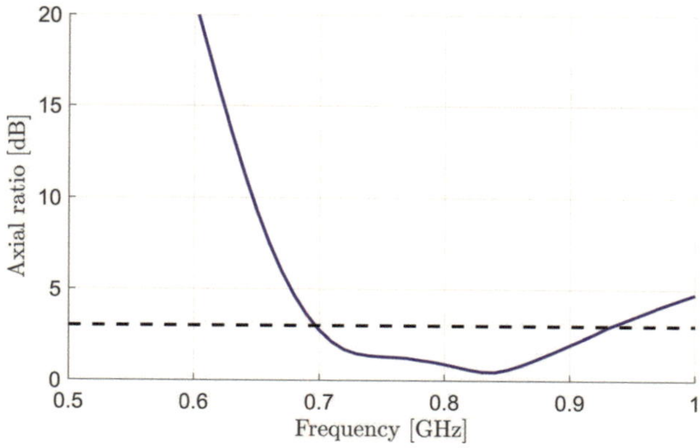

Figure 4.
Axial ratio vs. frequency of the spiral element.

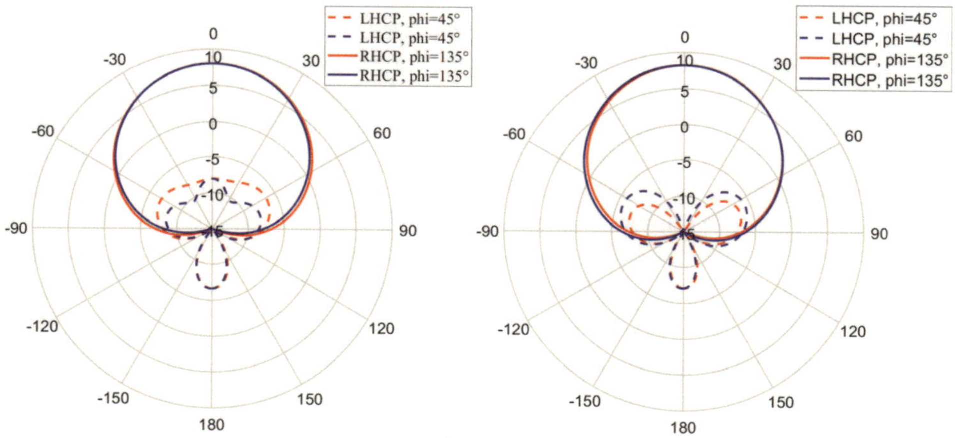

Figure 5.
Radiation patterns of the spiral element at 700 MHz (left) and 800 MHz (right).

The design provides a pretty high symmetry in the radiation patterns at all frequencies of the MIMO band and at all ϕ cut planes from 0 to 180°. Examples at 700 MHz and 800 MHz are plotted in **Figure 5** for two cuts for simplicity. The radiation patterns provide pure CP with a Cross-Polar Discrimination (XPD) at broadside above 18 dB and Side Lobe Level (SLL) greater than 26 dBi with almost constant HPBW of about 72°.

2.1.2 Dual-polarization spiral

In this section, a single four-arm spiral (**Figure 6**) is presented where the dual circular polarization can be obtained from a stand-alone antenna through the use of four arms fed by sequential feeding [24]. So the four arms are fed with 0, 90, 180, and 270° for LHCP and 0, 270, 180, and 90° for RHCP polarization [25]. It is true that this method dramatically affects the spiral BW; however, the other direct methods have other main limitations. For example, four spirals can be used to provide ultrawideband where each of two spirals is responsible for one polarization [26, 27].

This can be used at the single-element level, while at the array level, especially for the massive MIMO, this is not acceptable. This is not only because of the big area (16 antennas are needed for each 4-element subarray) but also because the symmetry between polarization is going to break and the high coupling between elements will strongly affect isolation between polarizations and isolation between subarrays.

Similar to the single-polarization spiral, a metallic object is needed for the unidirectional radiation patterns; a metallic cavity [25] is used here, which has the same rule as the planar reflector but with the addition of better isolation, which is preferred when the element is used in an array configuration. Also, two absorbers are included with dielectric constants $\epsilon_r = 1.1$ and 2 and loss tangents of 0.25 and 0.5 for the top/middle and bottom layers, respectively. **Figure 7** confirms the broadband behavior of the antenna from 680 MHz to 1 GHz, with almost the same behavior for both polarization due to the structural symmetry. However, another important issue for this design is the XPD, which usually is so high at boresight; for example, in [28], it shows that a poor XPD lower than 7 dB is observed at boresight for some frequencies inside the matching bandwidth.

In this design, to avoid these limitations and have pure CP radiation patterns with a flat gain (of about 9 dBi), as in **Figure 8**, the size of the single spiral needs to be enlarged to have a diameter of 400 mm and a cavity height of 140 mm. Also, it presents stable radiation patterns in the whole band as depicted in **Figure 8**, with an HPBW going up to 86 at the lower frequency band. Despite all the previous advantages that the solutions provide (broadband, symmetry, high/stable gain, clean radiation patterns, and pure double polarization purity), the extreme size of the element limits its use for massive MIMO base stations. Therefore in the next section, instead of the self-complementary broadband solution, resonating antennas will be investigated.

2.2 L-feed differential patch element

Contrary to the previous section, in the current one, a single patch antenna excited with 4 L- differential feeds [29, 30] is investigated (**Figure 9**). It can provide directly the dual slant linear polarization $(\pm 45^{\circ})$ by controlling the phases of the four feeds

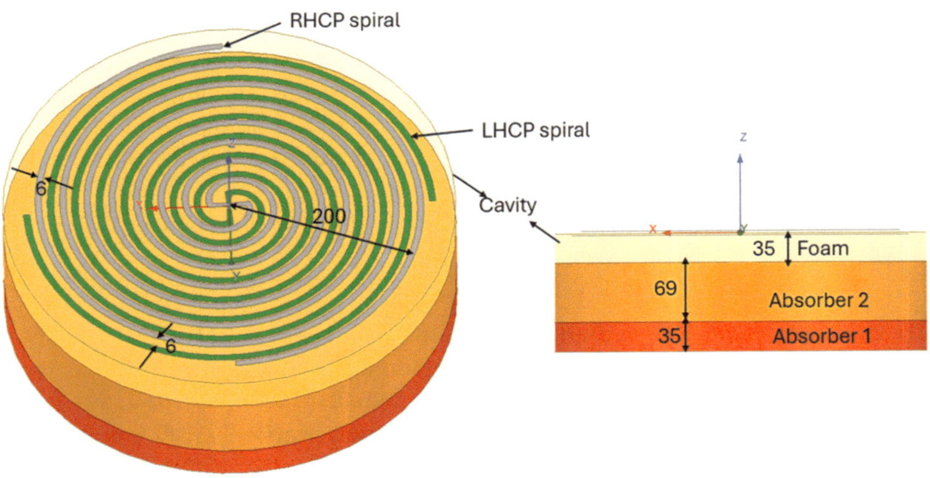

Figure 6.
Dual-polarized spiral element with dimensions in mm (left) perspective view (right) side view.

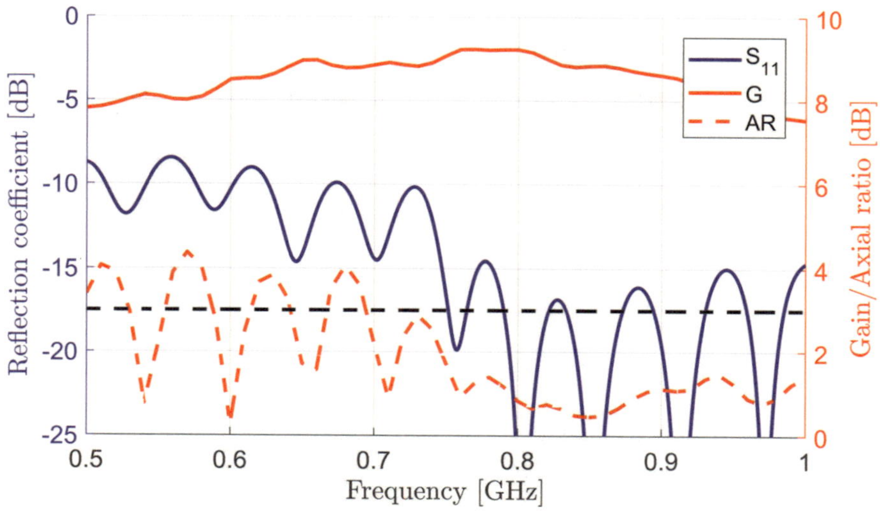

Figure 7.
S_{11} of dual-polarized spiral.

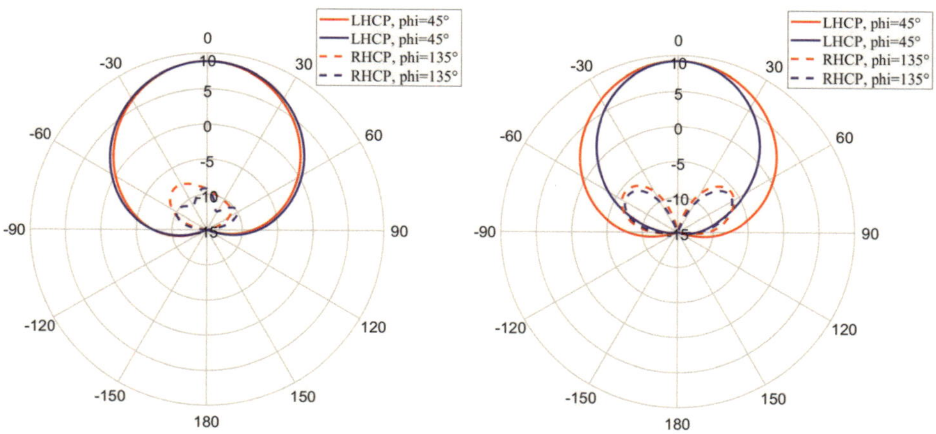

Figure 8.
Radiation patterns of dual-polarized spiral element at 700 MHz (left) and 800 MHz (right).

(X1, Y1, X2, and Y2) so feeding with amplitudes of 1, 1, 1, and 1 and phases of 0, 180, 0, and 180° respectively for one polarization and 1, 1, 1, and 1 and phases of 0, 0, 180, and 180°, for the other orthogonal polarization. The patch is printed on Arlon AD410 with a medium relative permittivity of $\epsilon_r = 4$ and $\tan\delta = 0.003$ and is included in a metallic cavity for enhancing the gain. The patch size fulfills the requirement of massive MIMO with a size of 100 mm × 100 mm which is corresponding to $0.21\lambda_{fmin} \times 0.21\lambda_{fmin}$ and covers broadband from 650 to 910 MHz which is, even more, wider than the required BW as presented in **Figure 10**.

Additionally, the design provides a high gain which has a linear increase with the frequency with a gain greater than 8 dBi in the band of interest. **Figure 11** represents the radiation patterns at two frequency samples corresponding to the minimum and maximum low-frequency base station band. This solution provides total symmetry radiation patterns at vertical and horizontal planes with perfect isolation between the

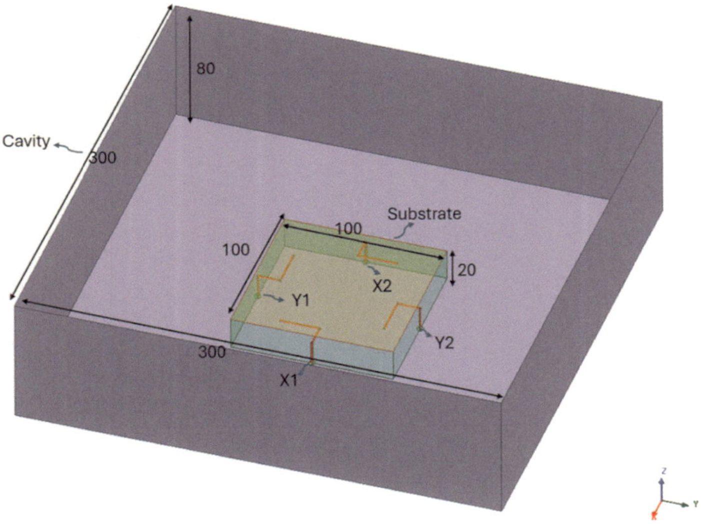

Figure 9.
Dual-polarized L-feed differential patch element with dimensions in mm.

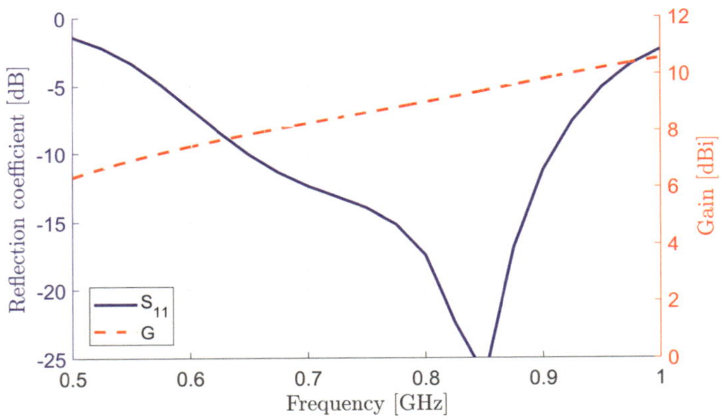

Figure 10.
S_{11} and gain versus frequency of the L-feed differential patch.

two polarizations of more than 50 dB in XPD and a front-to-back ratio above 16 dB. Also, due to the high gain, the solution has a narrow HPBW lower than 71° which facilitates obtaining a lower HPBW for the subarray. This element outperforms the dual-polarization spiral element, but the big size of the cavity of 300 mm × 300 mm × 80 mm makes it hard to be integrated at the array level. A single cavity can indeed be used to cover the complete subarray instead of a separate cavity for each element; however, this degrades the performance in terms of radiation patterns. Therefore, another element will be studied in the next section.

2.3 Differential metallic cap-loaded multi-layer DRA element

As we can conclude from the previous elements, the main challenge for massive MIMO base stations at low frequency is how to make the dual-polarization element

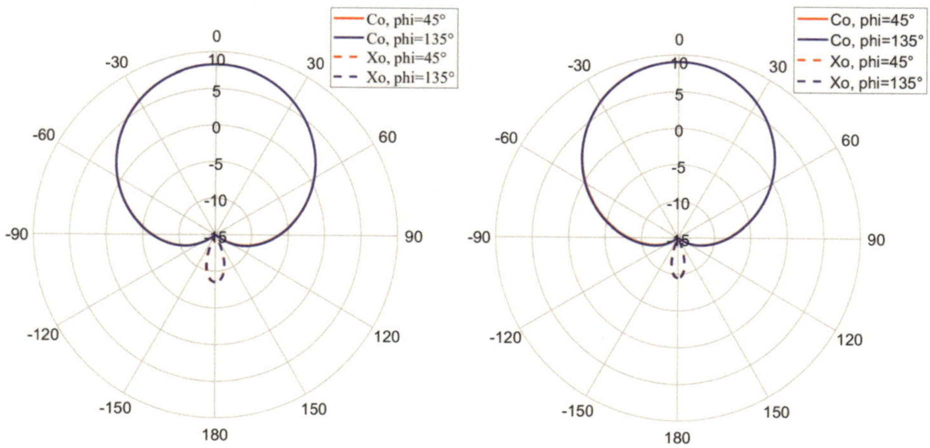

Figure 11.
Radiation patterns of dual-polarized L-feed differential patch at (a) 700 MHz and (b) 800 MHz.

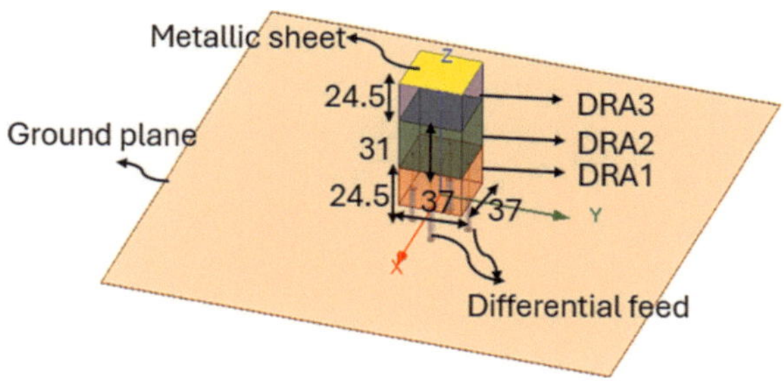

Figure 12.
Dual-polarized differential metallic cap-loaded multi-layer DRA element with dimensions in mm.

compact (lower than $0.23\lambda_{fmin} \times 0.23\lambda_{fmin}$) with keeping the broadband, high gain, good radiation patterns antenna performances. From the perspective of miniaturized and compact antennas, DRA can be a solution to that goal whenever it overcomes its inherent associated problems such as its extremely low bandwidth and gain performance that makes it less attractive for many broadband applications. It is well known that DRA can provide a very compact design by using high permittivity. Materials with very high permittivity up to $\epsilon_r = 140$ are commercially available [31]. For example, high values ($\epsilon_r > 35$) could provide elements with compact lengths smaller than 0.1λ, however, at the same time, the bandwidth is usually very small (lower than 5%) [12]. Hence bandwidth enhancement techniques are needed, several methods, can be used to offer broadband DRA, for example, tapered strip excitation [32] enables broadband but the gain is small, also, stair-shaped multi-layer [33], although it requires an increase of the size in one layer which is not preferable here.

After studying different broadband enhancement methods and their effects on other antenna parameters, multi-layer DRA with a fixed size and multiple materials [34] provides the best performance. It is based on exciting different modes and controlling them to form a broadband solution. However, the bandwidth is still not wide enough; therefore, the addition of a metallic cap [35] covering the top layer is proposed, as presented in **Figure 12**. The metallic cap at the top almost cancels the

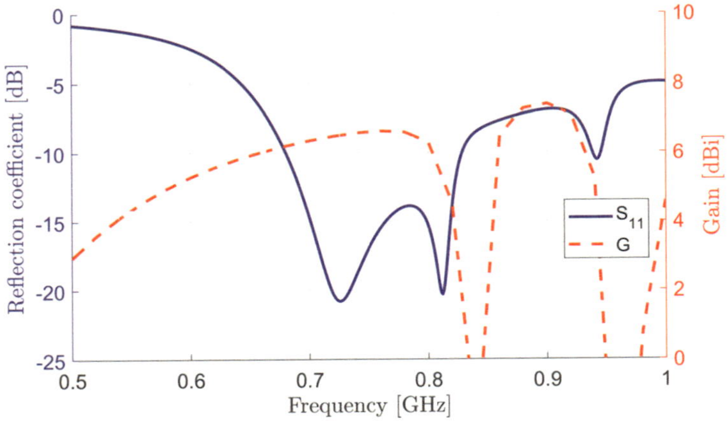

Figure 13.
S_{11} and gain of the differential metallic cap-loaded multi-layer DRA.

frequency dependence of the excited mode on the resonator height, making broadening BW achievable by increasing the height. Finally, differential feeding is used which provides higher BW compared to normal coaxial or microstrip feeding.

The reflection coefficient is depicted in **Figure 13**, and it shows that this solution covers from 680 to 830 MHz (20%), which is slightly larger than the band under interest. Additionally, it offers a flat gain of 6.5 dBi with an almost constant HPBW of $86 \pm 2^\circ$. This is achieved with a very compact dielectric element with a total size of 37 mm × 37 mm × 80 mm $\left(0.08\lambda_{fmin} \times 0.08\lambda_{fmin} \times 0.18\lambda_{fmin}\right)$. The DRA element consists of three blocks whose bottom and top layers are made of material with $\epsilon_r = 35$ while the middle layer has a permittivity of $\epsilon_r = 50$. Both materials are available *via* several dielectric resonator suppliers [31]. Compact DRA designs can be achieved with lower relative permittivity, but to preserve good performance, you need to modify the shape by using more complex shapes [36] or changing aspect ratios [37].

Similar to the L-feed differential patch, this proposed element provides perfectly symmetric patterns for both dual $\pm 45^\circ$ polarization and for all ϕ cut planes. As an example, the radiation patterns at one cut plane are plotted for 700 and 800 MHz in **Figure 14**. It has pure polarization discrimination with an XPD greater than 50 dB and a good FBR level above 11 dB.

To summarize this section, **Table 1** represents the main remarks for the discussed radiating elements.

Element	Polarization	Size $\left(\lambda_{fmin} \times \lambda_{fmin}\right)$	BW	Gain (dBi)	HPBW ($^\circ$)
Planar spiral 1	Single CP	0.19x0.19	700:920	8.0	72
Planar spiral 2	Dual CP	0.91x0.91	680:1000	7.3	68
L-feed differential patch	Dual slant	0.22x0.22	650:910	8.6	69
Cap-loaded multi-layer DRA	Dual slant	0.08x0.08	680:830	6.5	86

Table 1.
Comparison between the different radiating elements.

3. MIMO antenna array

From the discussion of Section 2 and considering the whole size restriction of the base station, compact elements are needed to implement massive MIMO and be able to fit eight subarrays; each subarray is a quadruplet consisting of four elements. Hence, the requirements for the subarray can be summarized as.

- Number of radiators: 64 (32× dual-polarization) elements in the form of 16 subarrays (8× dual-polarization) which each consist of 4 elements.
- Size: each subarray is $0.44\lambda_{fmin} \times 1.17\lambda_{fmin}$ which is corresponding to 187 mm × 500 mm for a minimum frequency of 700 MHz.
- HPBW: Smaller vertical HPBW of about 30 to 35°.
- Isolation: good isolation between subarrays lower than 20 dB.

The rest of this section is devoted to arrays based on the mentioned radiating elements; however, due to the large size of the cavity required for the L-feed differential patch element, this option will be discarded, and more focus will be directed toward the other resonating compact element (differential metallic cap-loaded multi-layer DRA element).

3.1 Spiral array

As discussed in Section 2.1, the dual-polarization spiral can not be used for MIMO applications due to its relatively large size (diameter of about $0.9\lambda_{fmin}$) where the size of the whole subarray is around $1\lambda_{fmin}$, and it should fit at least four elements to fulfill MIMO requirements from the point of view of HPBW. This is in addition to the requirement of having a horizontal size of $0.44\lambda_{fmin}$ to be able to support eight subarrays, besides some gaps for having good isolation between the subarrays. A

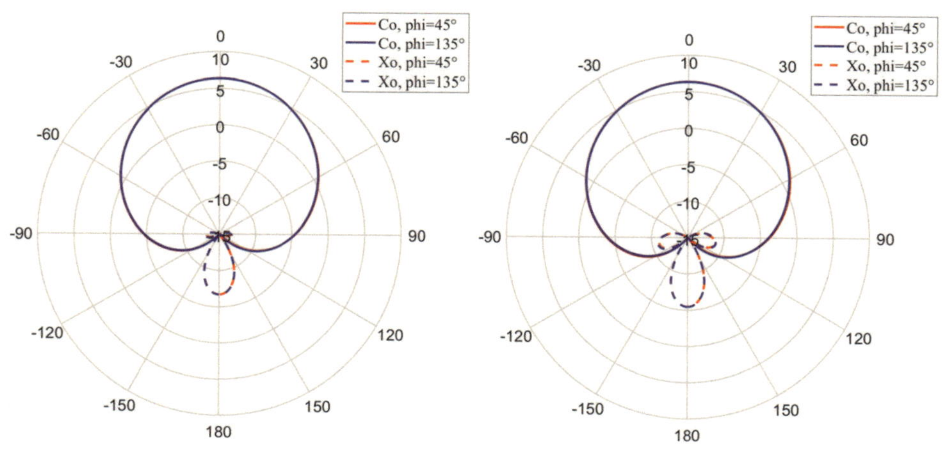

Figure 14.
Radiation patterns of dual-polarized differential metallic cap-loaded multi-layer DRA at 700 MHz (right) and 800 MHz (left).

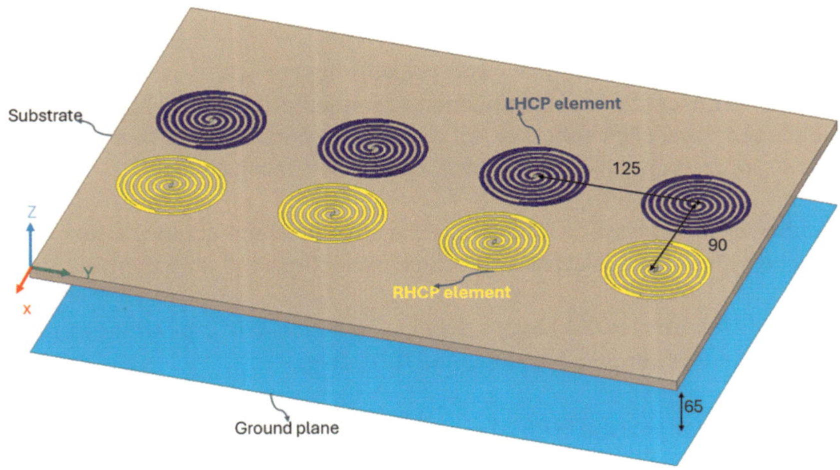

Figure 15.
A dual-polarized subarray based on single-polarization spiral element with dimensions in mm.

possible solution to overcome the first limitation is to have a separated subarray for each polarization based on the compact single-polarization spiral (Section 2.1.1), as presented in **Figure 15**.

Due to the mutual coupling between the spirals, which almost touch each other, the subarray response had a degradation in its bandwidth; however, it still covers the required band, as shown in **Figure 16**. It also enables having four radiators per subarray without breaking the symmetry. However, this is coming under the cost of failure to satisfy the second limitation for having a total number of eight subarrays per base station. In addition to the complex required feeding which consists of several baluns and circular-to-linear polarization converters. Hence in the following section, compact solutions will be studied to be able to get the required massive number of radiating elements of 16 subarrays (8× dual polarized).

3.2 DRA array

3.2.1 N = 4 DRA arrays

In this section, an array based on the proposed differential metallic cap-loaded multi-layer DRA of Section 2.3 will be analyzed. A truncated version of the complete array will be considered, consisting of three dual-polarized subarrays (instead of eight in the full version), as presented in **Figure 17**. To completely fulfill the massive MIMO requirements in terms of size, the spacing between elements is (125 mm) to fit the four elements on the 500 mm vertical side ($1.17\lambda_{700MHz}$), while the subarrays are set 187.5 mm apart from each other to fit the eight subarrays in the required size of 1500 mm ($1.17\lambda_{700MHz}$) for the horizontal side. With this truncated version, the complete effects in the array can be studied such as the performance of edge subarrays and mutual coupling for the internal subarrays without the need to run the extensive simulation for the complete MIMO system. Also, this enables us to measure isolation between neighbor subarrays in terms of first and second-neighbor interactions while no need to test for other farther neighbors as their coupling will be negligible.

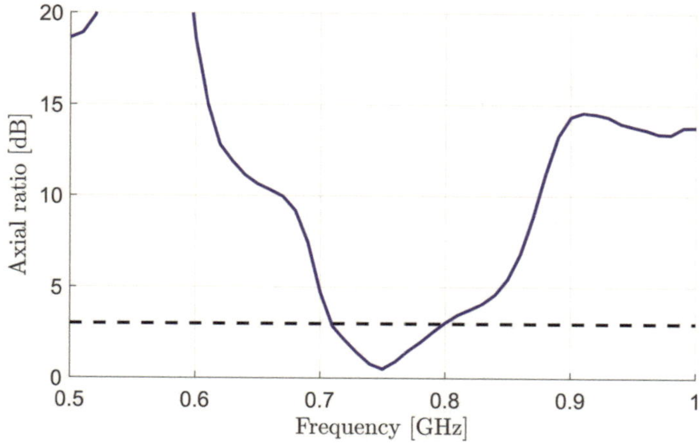

Figure 16.
Axial ratio of the dual-polarized spiral subarray.

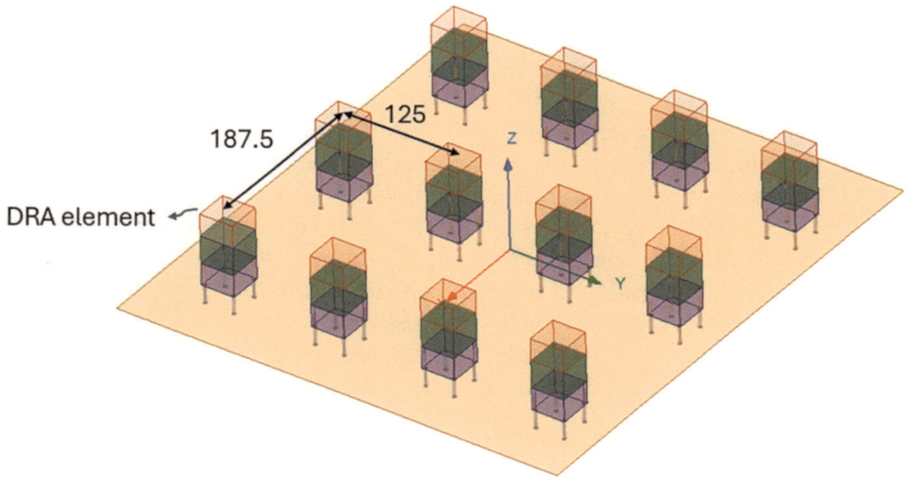

Figure 17.
A dual-polarized three quadruplet subarrays based on differential metallic cap-loaded multi-layer DRA.

For each polarization, the two differential feedings are connected through a balun to ensure the same amplitude with a 180° phase shift between them. Then, the output of the balun is connected to one leg of 4×1 combiner/power divider. The same is repeated for each element, then the output of the combiner/power divider forms the final port for this subarray at this polarization. So, in this truncated version, the full-wave simulation for the 12 dual-polarized elements is performed in HFSS using 48 separated ports, then the 48×48 s-parameters are exported in the form of *.s48p file. Finally, this file is imported to Applied Wave Research (AWR) microwave Office software [38], where the 48 ports are connected through 24 baluns, then six 4×1 combiners/power dividers to provide the final six output ports where each subarray has two ports one for -45° and other for $+45^{\circ}$ polarization. It is worth mentioning that the losses for this corporate feeding are about 1*to*2 dB; however, as the focus here is on the radiator itself, ideal feeding networks are used in AWR. Such losses are expected for other feeding methods such as Dielectric Image Guide (DIG) [39, 40]; however, it can be avoided by eliminating the combiners/power dividers by using other linear DRA array feeding techniques such as standing-wave [41], Substrate Integrated Waveguide (SIW) [42], or adjusting other high-efficiency broadband feedings for DRA element to be used in the array scale such as waveguides [43, 44] or cavity-backed slot [45].

Figure 18 presents the reflection coefficients S_{ii} of the three subarrays, for simplicity, the response of one of the external subarrays (ports 5 and 6) is omitted as it is symmetric to the other external ones (ports 1 and 2). It shows a high level of coherence between external and internal subarrays for both polarizations with a good matching level below −14 dB in the band of interest from 700 to 800 MHz.

The isolation between polarizations in the same subarray is below −16 and − 13.5 dB for internal and external subarrays, respectively, at the lower frequencies (**Figure 19**), with an enhancement at higher ones. Indeed, the isolation value is relatively weak which is mainly because of the high mutual coupling between polarizations due to their very small spatial separation (lower than $\lambda \div 10$ in the whole band). On the other hand, the isolation between subarrays (which is the more relevant for the MIMO systems) is excellent being lower than −20 dB in the whole band. Also, for the second-neighbor interaction such as S_{15} and S_{16}, the level is already negligible which confirms that there is no need to study farther neighbors, and three subarrays are enough to check the performance for the complete array without the need to test the full eight subarrays.

From the point of view of radiation characteristics, **Figure 20** shows the radiation patterns at the minimum, center, and maximum frequencies along with the gain as a function of the frequency of both co-polar and cross-polar components. For simplicity and due to the high symmetry, the results are shown for a single polarization and a single subarray (the middle one). The blue and red curves represent the radiation patterns at the vertical "direction of subarray" and horizontal planes, respectively, while solid and dashed lines mark the co-polar and cross-polar components, respectively.

Although the proposed element has almost constant HPBW as was presented in **Figure 14**, the array's HPBW at the vertical plane decreases with the increase of frequency

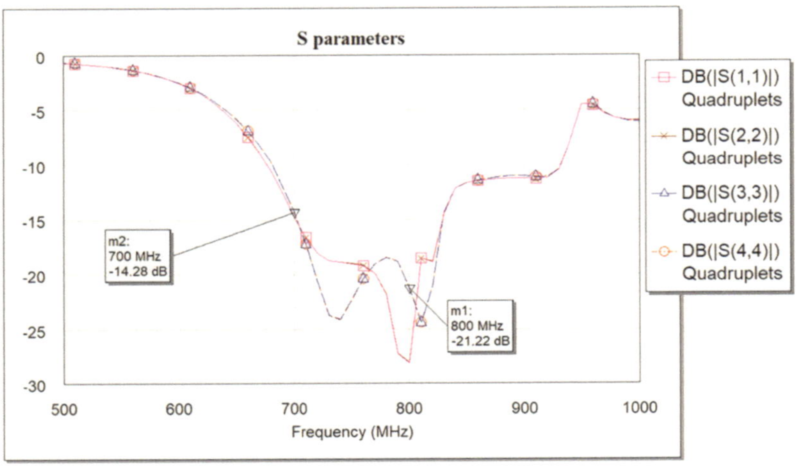

Figure 18.
Reflection coefficients S_{ii} of the three 4-element DRA subarrays.

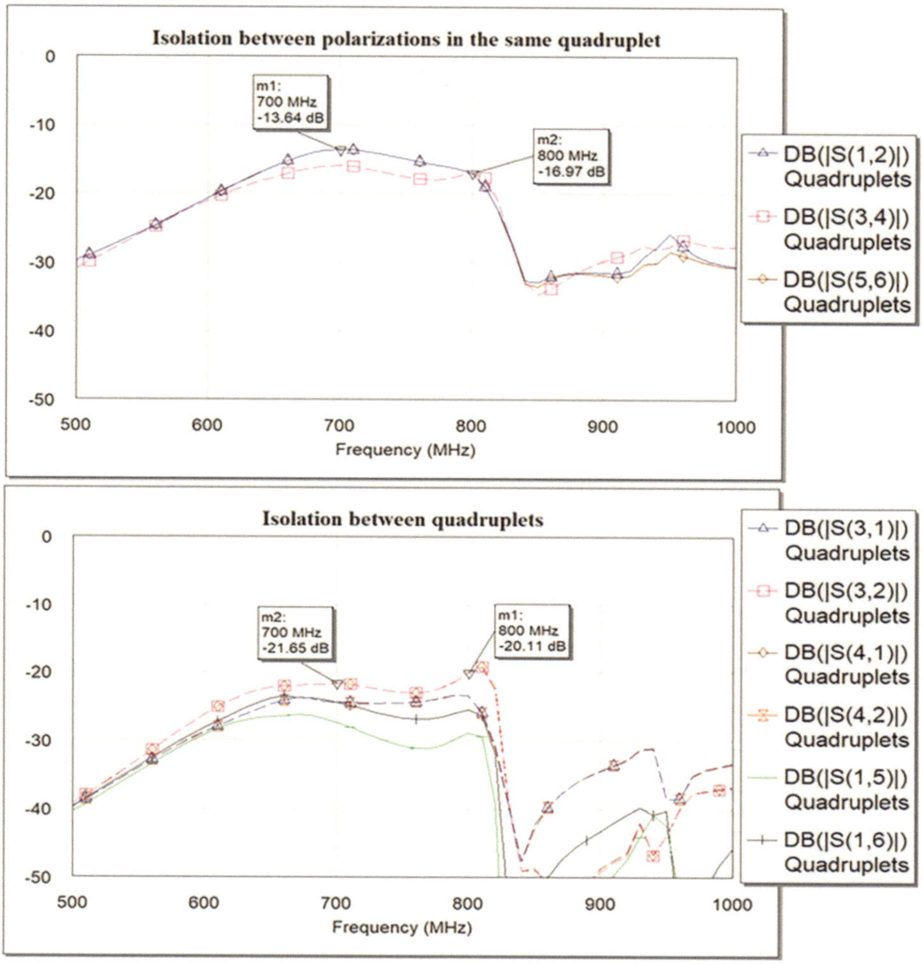

Figure 19.
Isolation S_{ij} between polarization S_{11} (top) and between different subarrays (bottom) of the three 4-element DRA subarrays.

due to the change of the element spacing in terms of electric length. The subarray reduces the vertical HPBW from 86 (for the element) to narrow HPBW of 32° and 36° at 800 and 750 MHz respectively. However, there is a slight increase to 38° at the lower band of 700 MHz. For the horizontal plane, the subarray has an almost fixed HPBW of $117 \pm 2°$. The subarray has an almost flat gain of 8.1 ± 0.3 dBi with XPD and FBR above 15 and 13 dB over the band under interest, respectively.

Overall, we can see that this array satisfies most of the massive MIMO requirements, especially in terms of the size where the 16T16R dual-polarized elements can fit in the base station. However, at the lowest frequencies, the HPBW has to be further reduced.

3.2.2 N = 8 DRA arrays

One of the main advantages of this compact element is the ability to further increase MIMO order by including more radiating elements without increasing the

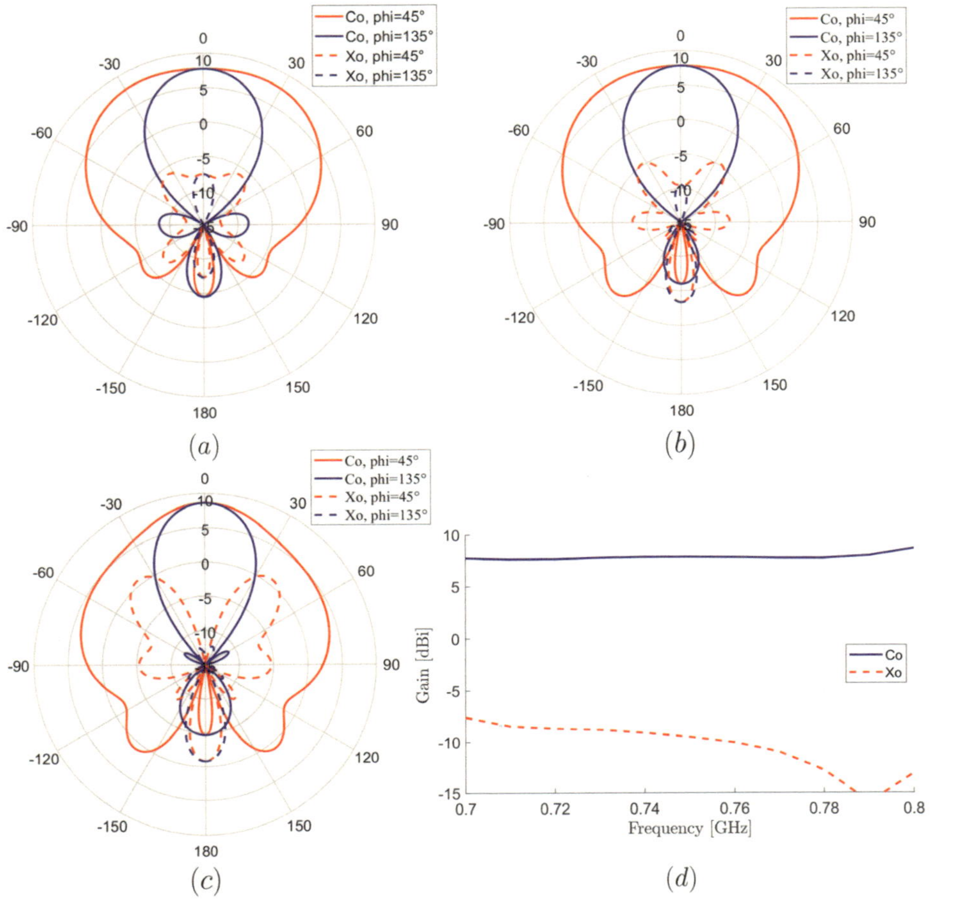

Figure 20.
Radiation patterns of the three 4-element DRA subarrays at (a) 700 MHz, (b) 750 MHz, (c) 800 MHz, and (d) gain vs. frequency.

overall size, as shown in **Figure 21**. It shows that eight dual-polarized elements can be fit in the spacing of a single subarray due to the compact size of the elements forming a base station with 8 × 8 dual-polarized elements.

Table 2 presents a comparison summary for different configurations with various excitations of the eight elements. The performance is compared from different perspectives including two of the main challenges for the massive MIMO which are low vertical HPBW and high gain. This, along with other standard merits, is discussed in this chapter. This demonstrates the possibility of this design providing unique properties such as the lowest HPBW lower than 30°, gain higher than 7 dBi, or a compromise between them.

3.2.3 Preliminary experimental data

A first prototype of the differential metallic cap-loaded multi-layer DRA element is manufactured. As presented, the DRA element consists of three blocks whose bottom and top layers are made of two materials with different materials. Here, the materials with $\epsilon_r = 35$ (for top and bottom layers) and the one for the middle layer (with a

Configuration								$HPBW_{Ver}$	$HPBW_{Hor}$	**Gain**	**FBR**	**XPD**
V0	**V1**	**V2**	**V3**	**V4**	**V5**	**V6**	**V7**	(°)	(°)	**(dBi)**	**(dB)**	**(dB)**
1	1	1	1	1	1	1	1	43	121	7.6	11.0	12.1
0	1	1	0	0	1	1	0	52	132	6.2	9.2	14.9
0	1	0	1	0	1	0	1	44	121	7.1	11.1	12.2
0	1	0	0	0	0	1	0	40	134	5.3	5.9	14.4
1	0	0	1	1	0	0	1	37	113	7.3	10.3	10.2
1	0	1	0	0	1	0	1	36	122	7.3	10.1	11.2
1	1	0	0	0	0	1	1	32	126	6.6	7.9	12.9
1	0	0	0	0	0	0	1	27	121	5.0	8.0	10.4

Table 2.
Different excitation configurations for 8-element subarrays.

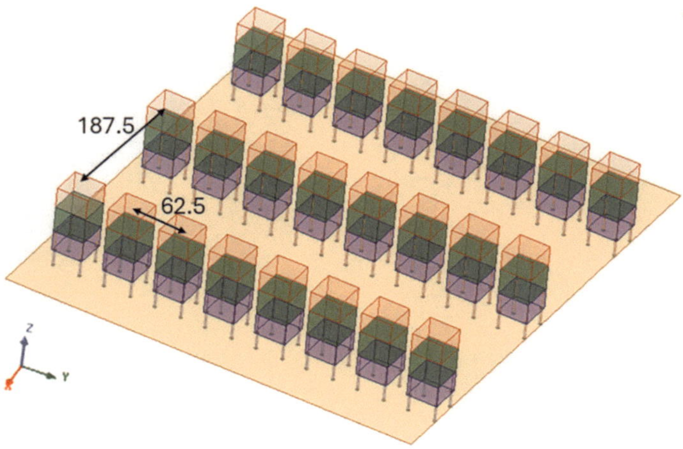

Figure 21.
Structure of a three 8-element subarrays.

permittivity of $\epsilon_r = 50$) are replaced by $\epsilon_r = 23$ and $\epsilon_r = 15$, respectively. This is because of the availability and easy handling of these materials [46]. Despite that the lower permittivity affects the size which, in turn, affects the performance, some adjuments for the radiator shape of the three layers [47] are proposed to compensate for this reduction and provide stable performance, as presented in **Figure 22**.

A comparison between the simulated and measured coefficient reflection of the fabricated prototype is shown in **Figure 23**. A good agreement is achieved with some discrepancies that can be due to the fabrication process which is done by injection molding followed in the second step by computer numerical control (CNC) machining [47]. The measured results show an impedance bandwidth of 13.15% (710–810 MHz). Furthermore, the measured radiation patterns at 700 and 800 MHz are plotted in **Figure 24**, which presents a high level of matching with XPD at broadside above 30 dB.

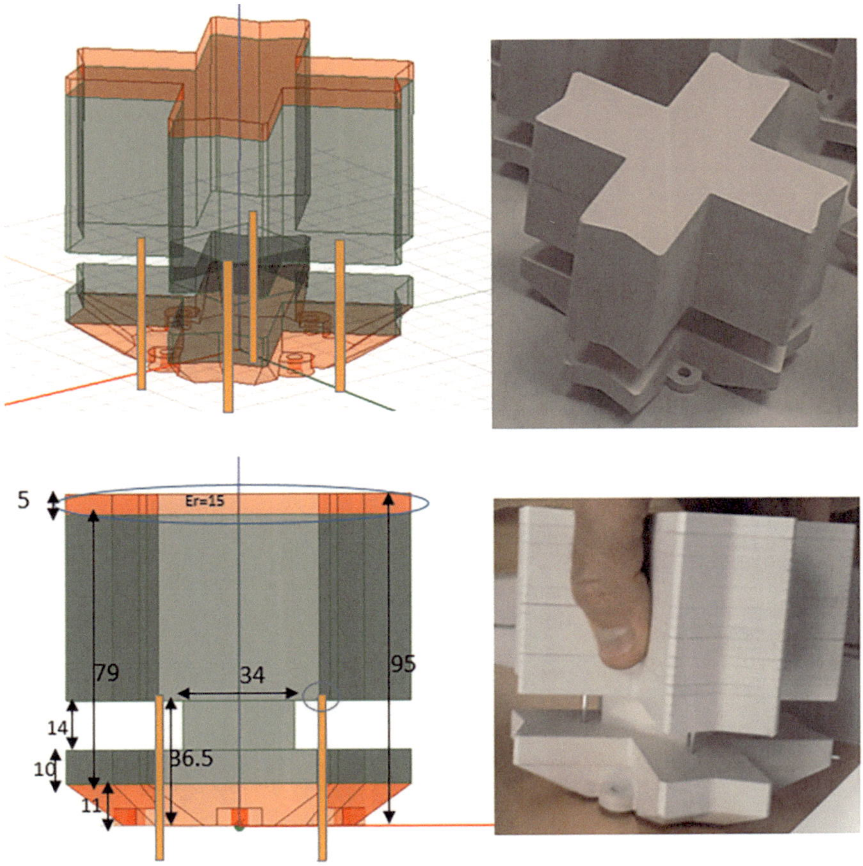

Figure 22.
Prototype of the differential metallic cap-loaded multi-layer DRA element.

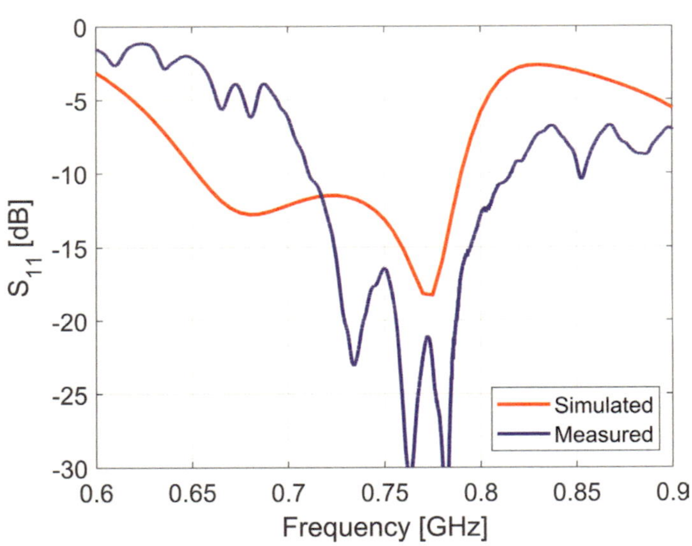

Figure 23.
Measured and simulated reflection coefficient results of the fabricated DRA antenna.

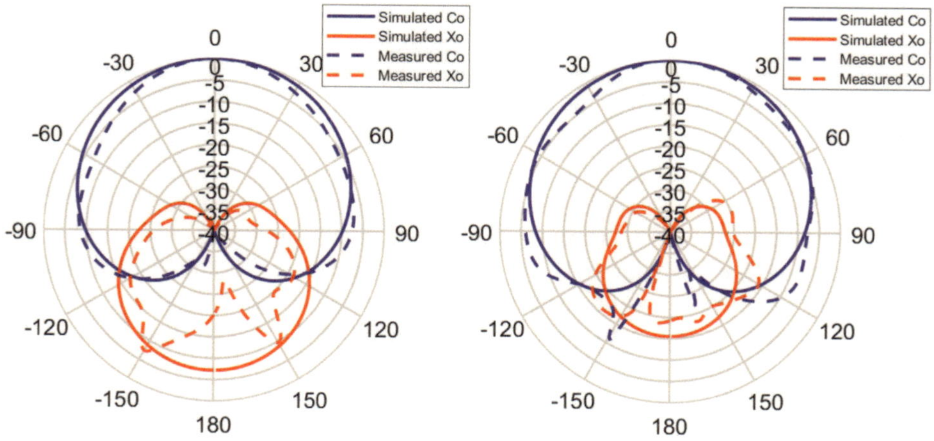

Figure 24.
Measured and simulated radiation patterns at 700 Mhz (left) and 800 MHz (right).

4. Conclusions

Different antenna elements are presented as suitable candidates for future massive MIMO base stations. The elements are based on a self-complementary planar spiral, a differential patch antenna, and a dielectric resonator antenna. Single-polarization planar spiral antenna provides excellent performance with a broadband impedance and CP bandwidth product from 700 MHz to 920 MHz. Also, it has an almost flat gain of about 7 dBi along the whole band. Then, the work is extended to provide a single spiral that could provide dual polarization. The design has the advantage of using only a single antenna and keeping the symmetry, which is important for isolation at the array level. However, to obtain stable radiation patterns at low frequency, the size of the spiral needs to significantly increase, which makes it unsuitable for massive dual polarization MIMO.

On the other hand, resonating antennas can provide a direct dual polarized which in its turn provides a compact design. An L-feed differential patch is shown which has a compact size of $\left(0.22\lambda_{fmin} \times 0.22\lambda_{fmin}\right)$ and covers a broad band from 650 to 910 MHz with a linearly-increasing gain from 8 to 10 dBi. Although the design is compact and its radiation patterns satisfy all massive MIMO requirements, a bigger cavity is necessary to achieve such performance which complicates the design of the array with a large number of radiating elements. Hence, another resonating element is studied, which is the dielectric resonator antenna (DRA). A differential metallic cap-loaded multi-layer DRA element is proposed, which seems a good candidate as it has a compact size of $\left(0.08\lambda_{fmin} \times 0.08\lambda_{fmin}\right)$ and still covers a wide enough band of about 20% from 680 to 830 MHz. This is in addition to a flat high gain of 6 dBi and symmetric radiation patterns with excellent properties.

An array version based on the self-complementary planar single-polarized spiral element is studied. The performance of the array is good; however, half of the space is lost as this solution needs two subarrays to represent a single dual-polarized. This limits the maximum number of subarrays to four subarrays instead of eight. Therefore, the focus is directed toward resonating elements where the differential metallic cap-loaded multi-layer DRA outperforms the L-feed differential patch element. Not only in terms of the size of the radiating element but also because the latter needs a relatively big cavity to have good performance. A truncated version of three subarrays is studied carefully, and each subarray consists of four of the proposed DRA elements. The array covers the complete band from 700 to 800 MHz with excellent levels of matching and isolation between subarrays below −14 and − 20 dB, respectively. Also, it has an almost flat gain of 8.1 ± 0.3 dBi with an average HPBW of 35°. This work was extended too to include the double number of elements to increase further the MIMO order as eight elements can be fit in each subarray, enabling the base station to carry 128 elements (64 × dual-polarized) and provide attractive adjustments of the array performance. This is along with several tradeoffs between high gain and low beamwidth, which are the main goals for future massive MIMO applications.

Acknowledgements

We want to thank HUAWEI Technologies Dusseldor GMBH, Germany for the productive meetings we had and the innovative ideas raised related to antenna design and fabrication methods. This work was supported by TSI-06000-2021-(128/125). Thanks to Ivan Bailon Ballesteros for his support in the simulations.

The work of Kerlos Atia Abdalmalak is supported by the Recualificacion-Margarita Salas grant (call of Universidad Carlos III de Madrid) financed by the Ministry of Science, Innovation, and Universities and the European Union-Next Generation EU and WiHEAT-CM-UC3M financed by Comunidad de Madrid. Also, he wants to acknowledge the grant provided through the 2023 IEEE Antennas and Propagation Society Fellowship (APSF) Program.

Conflict of interest

The authors declare no conflict of interest.

Abbreviations

AWR	applied wave research
BW	bandwidth
CP	circular polarization
CSRR	complementary split ring resonator
CST	computer simulation technology
DIG	dielectric image guide
DRA	dielectric resonator antenna
FBR	front-to-back ratio
HFSS	high-frequency simulation software
HPBW	half-power beamwidth
IoT	internet of things
LHCP	left-hand circular polarization
MIMO	multiple-input and multiple-output
RHCP	right-hand circular polarization
SIW	substrate integrated waveguide

SLL	side lobe level
TRX	transceivers
XPD	cross-polar discrimination

Author details

Kerlos Atia Abdalmalak[1,2]*, Ahmed El Yousfi[1] and Daniel Segovia Vargas[1]

1 Department of Signal Theory and Communications, Carlos III University of Madrid, Madrid, Spain

2 Department of Electrical Engineering, Aswan University, Aswan, Egypt

*Address all correspondence to: kabdalma@ing.uc3m.es

References

[1] Aerts S, Verloock L, Van Den Bossche M, Colombi D, Martens L, Tornevik C, et al. In-situ measurement methodology for the assessment of 5G NR massive MIMO Base station exposure at Sub-6 GHz frequencies. IEEE Access. 2019;7:184658-184667

[2] Larsson EG, Marzetta TL, Ngo HQ, Yang H. Antenna count for massive MIMO: 1.9 GHz vs. 60 GHz. IEEE Communications Magazine. 2018;**56**(9): 132-137

[3] Ma R, Gao Y, Cuthbert L, Zeng Q. Antipodal linearly tapered slot antenna array for millimeter-wave base station in massive MIMO systems. In: 2014 IEEE Antennas and Propagation Society International Symposium (APSURSI). United States: IEEE; 2014. pp. 1121-1122

[4] Qin Y, Zhang L, Mao CX, Zhu H. A compact wideband antenna with suppressed mutual coupling for 5G MIMO applications. IEEE Antennas and Wireless Propagation Letters. 2023; **22**(4):938-942

[5] Zhou Z, Wei Z, Tang Z, Yin Y. Design and analysis of a wideband multiple-microstrip dipole antenna with high isolation. IEEE Antennas and Wireless Propagation Letters. 2019;**18**(4):722-726

[6] Ryu KS, Kishk AA. Wideband dual-polarized microstrip patch excited by hook shaped probes. IEEE Transactions on Antennas and Propagation. 2008; **56**(12):3645-3649

[7] Jin Y, Du Z. Broadband dual-polarized F-probe fed stacked patch antenna for base stations. IEEE Antennas and Wireless Propagation Letters. 2015; **14**:1121-1124

[8] Cui Y, Niu Y, Qin Y, Li R. A new high-isolation broadband flush-mountable dual-polarized antenna. IEEE Transactions on Antennas and Propagation. 2018;**66**(12):7342-7347

[9] Lamkaddem A, Yousfi AE, Abdalmalak KA, Posadas VG, Segovia-Vargas D. Circularly polarized miniaturized implantable antenna for leadless pacemaker devices. IEEE Transactions on Antennas and Propagation. 2022;**70**(8):6423-6432

[10] Chouhan S, Panda DK, Gupta M, Singhal S. Meander line MIMO antenna for 5.8 GHz WLAN application. International Journal of RF and Microwave Computer-Aided Engineering. 2018;**28**(4):e21222

[11] El Yousfi A, Lamkaddem A, Abdalmalak KA, Segovia-Vargas D. A miniaturized triple-band and dual-polarized monopole antenna based on a CSRR perturbed ground plane. IEEE Access. 2021;**9**:164292-164299

[12] Petosa A, Ittipiboon A. Dielectric resonator antennas: A historical review and the current state of the art. IEEE Antennas and Propagation Magazine. 2010;**52**(5):91-116

[13] Tang H, Chen JX, Yang WW, Zhou LH, Li W. Differential dual-band dual-polarized dielectric resonator antenna. IEEE Transactions on Antennas and Propagation. 2017;**65**(2):855-860

[14] Wang XY, Tang SC, Yang LL, Chen JX. Differential-fed dual-polarized dielectric patch antenna with gain enhancement based on higher order modes. IEEE Antennas and Wireless Propagation Letters. 2020;**19**(3):502-506

[15] Zhang ZY, Wu KL. A wideband dual-polarized dielectric magnetoelectric

dipole antenna. IEEE Transactions on Antennas and Propagation. 2018;**66**(10): 5590-5595

[16] Wang Z, Dong Y, Peng Z, Hong W. Hybrid metasurface, dielectric resonator, low-cost, wide-angle beam-scanning antenna for 5G base station application. IEEE Transactions on Antennas and Propagation. 2022;**70**(9):7646-7658

[17] Zhao SK, Liu NW, Chen Q, Fu G, Chen XP. A low-profile dielectric resonator antenna with compact-size and wide bandwidth by using metasurface. IEEE Access. 2021;**9**: 29819-29826

[18] Das K. Conversion from linear to circular polarization in FPGA in real time [Doctoral dissertation]. Germany: University of Cologne; 2013 Available from: https://kups.ub.uni-koeln.de/5201/

[19] Ripoche O, Aubert H, Bellion A, Potier P, Pouliguen P. Spiral antenna miniaturization in very high frequency band. In: 2012 15 International Symposium on Antenna Technology and Applied Electromagnetics. United States: IEEE; 2012. pp. 1-5

[20] Computer Simulation Technology, CST. 2024. Available from: https://www.cst.com/

[21] ANSYS. Simulation driven product development, HFSS. Available from: https://www.ansys.com/

[22] Abdalmalak KA, Santamaría Botello G, Romano SL, García Muñoz LE, Vargas DS. An updated version of the Dyson conical quad-spiral Array (DYQSA) feed system for VGOS applications. In: 2017 IEEE International Symposium on Antennas and Propagation and USNC/URSI National Radio Science Meeting. United States: IEEE; 2017. pp. 1539-1540

[23] Abdalmalak KA, Romano SL, Garcia E, Lamperez AG, Martinez FJH, Palma MS, et al. Radio astronomy ultra wideband receiver covering the 2-14 GHz frequency band for VGOS applications. In: 2016 10th European Conference on Antennas and Propagation (EuCAP). United States: IEEE; 2016. pp. 1-5

[24] Fang HR, Serhir M, Guinvarc'h R, Mouthaan K. Enhanced dual-circular polarised four-arm Archimedean spiral antenna with low-profile cavity backing. IET Microwaves, Antennas and Propagation. 2015;**9**(12):1260-1266

[25] Herrera-Martin JM, Gonzlez-Posadas V, Segovia-Vargas D. 183GHz cavity-backed spiral radiometer antenna with dual circular polarization for earth observation applications. In: 2022 16th European Conference on Antennas and Propagation (EuCAP). United States: IEEE; 2022. pp. 1-5

[26] Abdalmalak KA, Santamaría Botello G, Llorente-Romano S, Rivera-Lavado A, Flygare J, Lopez Fernándezs JA, et al. Ultrawideband conical log-spiral circularly polarized feed for radio astronomy. IEEE Transactions on Antennas and Propagation. 2020;**68**(3):1995-2007

[27] Abdalmalak, Dawoud KA. Analysis and design of antennas and radiometers for radio astronomy applications in microwave, Mm-wave, and THz bands [Doctoral dissertation]. Spain: Universidad Carlos III de Madrid; 2022. Available from: http://hdl.handle.net/10016/35126

[28] Baard C, Liu Y, Nikolova N. Ultra-wideband low-cost high-efficiency cavity-backed compound spiral antenna. Electronics. 2020;**9**(9):1399

[29] Yang X, Ji Y, Ge L, Zeng X, Wu Y, Liu Y. A dual-band radiation-

differentiated patch antenna for future wireless scenes. IEEE Antennas and Wireless Propagation Letters. 2020; **19**(6):1007-1011

[30] Li P, Luk KM, Lau KL. A dual-feed dual-band L-probe patch antenna. IEEE Transactions on Antennas and Propagation. 2005;**53**(7):2321-2323

[31] Luk KM, Leung KW. Dielectric Resonator Antennas. Antennas S. United Kingdom: Research Studies Press; 2003

[32] Khalily M, Rahim MKA, Kishk AA. Bandwidth enhancement and radiation characteristics improvement of rectangular dielectric resonator antenna. IEEE Antennas and Wireless Propagation Letters. 2011;**10**:393-395

[33] Huang W, Kishk AA. Compact dielectric resonator antenna for microwave breast cancer detection. IET Microwaves, Antennas and Propagation. 2009;**3**(4):638-644

[34] Huang W, W, Kishk AA. Compact wideband multi-layer cylindrical dielectric resonator antennas. IET Microwaves, Antennas and Propagation. 2007;**1**(7):998-1005

[35] Shaik LA, Saha C, Arora S, Das S, Siddiqui JY, Iyer AK. Bandwidth control of cylindrical ring dielectric resonator antennas using metallic cap and sleeve loading. IET Microwaves, Antennas and Propagation. 2017;**11**(12): 1742-1747

[36] Yousfi AE, Abdalmalak KA, Lamkaddem A, Vargas DS. Miniaturized broadband dual-polarized dielectric resonator antenna using characteristic modes. In: 2023 17th European Conference on Antennas and Propagation (EuCAP). United States: IEEE; 2023. pp. 1-4

[37] Abdalmalak KA, Santamaría-Botello G, Lee CS, Rivera-Lavado A, García-Castillo LE, Segovia-Vargas D, et al. Microwave radiation coupling into a WGM resonator for a high-photonic-efficiency nonlinear receiver. In: 2018 48th European Microwave Conference (EuMC). United States: IEEE; 2018. pp. 781-784

[38] Applied Wave Research, AWR. 2024. Available from: https://www.cadence.com/

[39] Al-Zoubi AS, Kishk AA, Glisson AW. A linear rectangular dielectric resonator antenna array fed by dielectric image guide with low cross polarization. IEEE Transactions on Antennas and Propagation. 2010;**58**(3):697-705

[40] Al-Zoubi AS, Kishk AA, Glisson AW. Aperture coupled rectangular dielectric resonator antenna Array fed by dielectric image guide. IEEE Transactions on Antennas and Propagation. 2009;**57**(8):2252-2259

[41] Abdalmalak KA, Althuwayb AA, Lee CS, Botello GS, Falcon-Gomez E, Garcia-Castillo LE, et al. Standing-wave feeding for high-gain linear dielectric resonator antenna (DRA) array. Sensors. 2022;**22**(8)

[42] Abdel-Wahab WM, Busuioc D, Safavi-Naeini S. Millimeter-wave high radiation efficiency planar waveguide series-fed dielectric resonator antenna (DRA) array: Analysis, design, and measurements. IEEE Transactions on Antennas and Propagation. 2011;**59**(8): 2834-2843

[43] Alsirhani K, Abdalmalak KA, Lee CS, Santamaria-Botello G, Segovia-Vargas D, Garcia-Muoz LE. Dielectric resonator antenna fed by tapered dielectric rod waveguide for 5G mm-

wave applications. In: 2020 IEEE International Symposium on Antennas and Propagation and North American Radio Science Meeting. United States: IEEE; 2020. pp. 149-150

[44] Eshrah IA, Kishk AA, Yakovlev AB, Glisson AW. Excitation of dielectric resonator antennas by a waveguide probe: Modeling technique and wide-band design. IEEE Transactions on Antennas and Propagation. 2005;**53**(3): 1028-1037

[45] Alsirhani KF, Abdalmalak KA, Lee CS, Althuwayb AA, Posadas VG, Munoz LEG. Enhancement of gain and bandwidth of cylindrical dielectric resonator antenna excited by cavity-backed slot. Alexandria Engineering Journal. 2024;**104**:480-489. Available from: https://www.sciencedirect.com/science/article/pii/S111001682400872X

[46] PREPERM Low Loss Dielectric Thermoplastics. Available from: https://www.avient.com/

[47] Yousfi AE, Abdalmalak KA, Lamkaddem A, Barrera AM, Biscontini B, Segovia-Vargas D. Miniaturized dual-polarized, high-gain, and wideband dielectric resonator antenna for low band massive MIMO applications. Progress in Electromagnetics Research. 2024;**179**: 101-111

Chapter 8

Sparse Polarimetric Array for MIMO System

Yaxing Yue, Guisheng Liao, Xin Yuan and Zhiguo Shi

Abstract

To enhance sensing or communication capabilities, the utilization of extremely large multiple-input multiple-output transceiver arrays (EL-MIMO-TAs) with high array degrees-of-freedom (DoFs) holds significant promise for future applications, such as integrated sensing and communications systems. However, the implementation of EL-MIMO-TAs may face the problem of mutual coupling. It also necessitates a substantial number of radio frequency chains, leading to undesirable hardware costs. Consequently, there has been a growing interest in sparse array designs aimed at reducing mutual coupling and hardware expenses while maintaining a constant array DoFs. Nevertheless, conventional sparse arrays primarily benefit from spatial DoFs alone. A shift toward the development of sparse polarimetric arrays has emerged, offering low mutual coupling, and enhanced spatial DoFs along with the ability to exploit the polarimetric characteristics of electromagnetic waves. This chapter explores different strategies for designing sparse polarimetric arrays, focusing on the sparsity of array element positions—specifically, non-uniform, uniform, and hybrid non-uniform and uniform sparsity. Additionally, it introduces a novel method for estimating multi-dimensional parameters based on the reconstructed covariance matrix through data fitting, emphasizing low computational complexity. A new beamformer in the spatial-polarimetric joint domain is also presented, showcasing its ability to suppress main-lobe interferences and improved beamforming performance from a sparse reconstruction perspective.

Keywords: enhanced degrees-of-freedom, MIMO system, reduced hardware costs, signal processing, sparse polarimetric array

1. Introduction

Multiple-input multiple-output (MIMO) systems are an important foundation for achieving sensing and communication [1–5]. In the future, high-performance sensing and communication will inevitably require large-scale antenna arrays as the transceiver units of MIMO systems [6]. However, large-scale antenna arrays correspond to large-scale radio frequency chains, which means a high computational burden for real implementations. While dense large-scale antenna arrays are beneficial for achieving communication functions, they may face the challenge of relatively limited coverage range. Additionally, when implementing sensing functions, they may encounter severe mutual coupling effects [7–9], thus affecting sensing effectiveness [10–12]. Furthermore, electromagnetic waves are vector signals, and failure to consider the polarization characteristics of signals will lead to loss of information dimensionality [13, 14]. Therefore, researching configurations of sparse polarimetric array with high degrees of freedom (DoFs), low mutual coupling, and polarization information processing capabilities is crucial for serving large coverage area communication, high-precision sensing, and even integrated communication and sensing systems [15].

Unlike the sparse scalar arrays where scarification is optimized only in terms of position [16, 17], sparse polarimetric arrays have more diverse configuration optimization methods due to their three characteristics: diverse polarization types (e.g., dipole or loop) of array elements [18], diverse pointing directions of array elements, and diverse positions of array elements. Among these three diverse characteristics, the position of array elements is the most important and obvious. In this section, based on this characteristic, we start from three ways of non-uniform sparsity, uniform sparsity, and mixed non-uniform and uniform sparsity in terms of array element positions, providing common structural optimization design methods for sparse polarimetric arrays and analyzing the advantages and disadvantages of various sparse design methods. Then, using a mixed design approach as an example in this chapter, we will introduce the signal model for sparse polarimetric arrays, a data fitting method to reconstruct the equivalent data for the uniform counterpart of sparse polarimetric arrays, an enhanced direction-of-arrival (DOA) estimation technique with increased DOFs in closed form, and a more powerful beamforming algorithm in joint spatial and polarimetric domain with improved performance.

Notations: $(\cdot)^*$, $(\cdot)^{\mathrm{T}}$, and $(\cdot)^{\mathrm{H}}$ mean the conjugate operator, transpose operator, and Hermitian operator, respectively. $*$, $\otimes$, and $\circledast$ means convolution, Kronecker, and Khatri-Rao product, respectively. $\mathrm{diag}(\cdot)$ and $\mathrm{blkdiag}(\cdot)$ denote the diagonal matrix operator and block-diagonal matrix operator, respectively. $\mathbf{I}_n$ and $\mathbf{O}_n$ are the identity matrix and zero matrix of the dimension $n \times n$, respectively. $\mathbf{1}_n$ represents the $n \times 1$ vector of ones. $\mathbf{O}_{n_1 \times n_2}$ means the zero matrix of the dimension $n_1 \times n_2$. $\mathbf{i}_n = \mathrm{vec}(\mathbf{I}_n)$ with $\mathrm{vec}(\cdot)$ denoting vectorization. $\Sigma\mathrm{diag}(\mathbf{X}, n)$ denotes the sum of the elements along the n-th diagonal of the matrix $\mathbf{X}$. $[\mathbf{x}]_n$ refers to the n-th element of the vector $\mathbf{x}$, whereas $[\mathbf{X}]_{n_1,n_2}$ denotes the (n_1, n_2)-th entry of the matrix $\mathbf{X}$. $|\cdot|$, $\|\cdot\|$, and $\|\cdot\|_{\mathrm{F}}$ are the modulus, Euclidean norm, and Frobenius norm, respectively. The n-dimensional complex-valued and real-valued vectors are denoted by $\mathbb{R}^n$ and $\mathbb{C}^n$, respectively. $\mathbb{N}^{n_1 \times n_2}$, $\mathbb{R}^{n_1 \times n_2}$, and $\mathbb{C}^{n_1 \times n_2}$ represent the $n_1 \times n_2$ matrices with natural numbers, real-valued numbers, and complex-valued numbers, respectively. $\mathrm{tr}(\cdot)$, $E(\cdot)$, $\angle(\cdot)$, and $\det(\cdot)$ stand for the trace, expectation, angle, and determinant, respectively.

2. Design methodologies for sparse polarimetric arrays

2.1 Design of sparse polarimetric arrays based on non-uniform sparsity

The existing research on non-uniform sparse polarimetric arrays is mainly based on the simple extension from common sparse scalar arrays, such as coprime array [19, 20], nested array (NA) [21], super-nested array [22], and augmented nested array (ANA) [23], to polarimetric arrays, and is mostly applied to one-dimensional DOA and polarization estimation [24]. For example, a sparse polarimetric array composed of NAs and ANAs was proposed in [25] (as shown in **Figure 1**), which not only

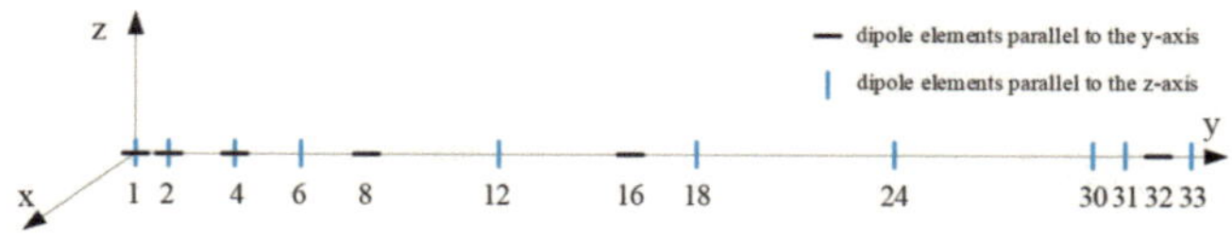

Figure 1.
Sparse polarimetric array composed of 6 NA dipoles (axial directions parallel to the y-axis) and 10 ANA dipoles (axial directions parallel to the z-axis).

achieves aperture extension but also avoids mutual coupling effects between array elements. The design concept of the above-mentioned sparse polarimetric array is as follows: (1) The ideal NA dipole elements parallel to the y-axis starts from the origin are sequentially placed along the y-axis, while the ANA dipole elements parallel to the z-axis are sequentially placed along the y-axis; (2) It is better to equalize the number of virtual NA dipole elements and ANA dipole elements as much as possible.

Although the array shown in **Figure 1** has the advantages mentioned above, its spatial aperture is only along the y-axis, and it only perceives polarimetric information using two kinds of polarimetric antennas, leading to a loss of potential for unambiguous estimation of two-dimensional (2D) DOA and polarization parameters. The nested electromagnetic vector antenna (EMVA) array [26] shown in **Figure 2**, although also having a spatial aperture only along the y-axis, can perceive polarimetric information using six kinds of polarimetric antennas. It can achieve unambiguous estimation of 2D DOA and polarization parameters by leveraging the "vector cross-product" property [27].

The design concept of the above array is to use EMVA elements as the basic units arranged at the NA positions. The first layer of the two-layer nested EMVA array shown in **Figure 2** consists of N_1 EMVA elements (with element spacing of d_y), and the second layer consists of N_2 EMVA elements (with element spacing of $(N_1+1)d_y$). In the figure, the red (black, blue) short dashed lines represent dipole elements whose axial directions are parallel to the x (y, z) axis, and the red (black, blue) (elliptical) circles represent magnetic loops whose normal directions are parallel to the x (y, z) axis.

Although the array shown in **Figure 2** has the potential for unambiguous estimation of multi-dimensional parameters, the compact non-orthogonal magnetic loops and dipoles in the array may lead to significant mutual coupling effects within the EMVA elements. To address this issue, in addition to self-calibration methods [7], existing literature has shown two potential solutions. One approach is stretching collocated EMVA elements into non-collocated ones to reduce the strong coupling effects caused by collocated configurations [28]. Another approach to address the mutual coupling effects within the EMVA elements involves stretching paired orthogonal magnetic loops and dipoles. This solution leverages the orthogonal electromagnetic fields of the magnetic loops and dipoles [29], resulting in smaller coupling effects. Furthermore, further stretching of the orthogonal magnetic loops and

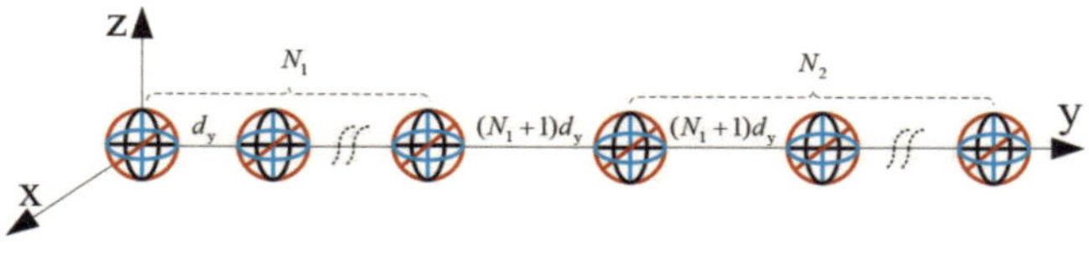

Figure 2.
Sparse polarimetric array composed of two-layer nested EMVA sub-arrays.

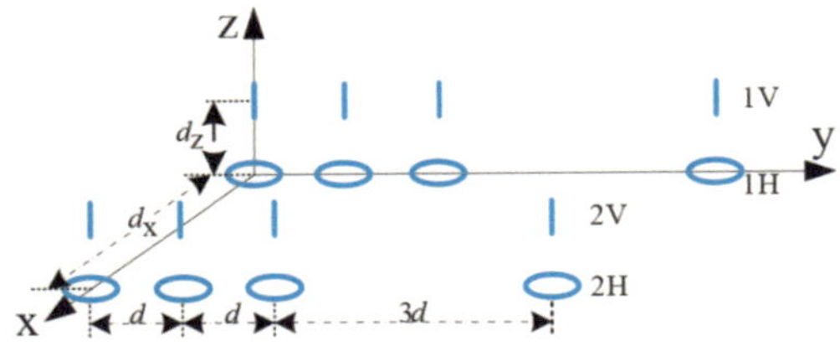

Figure 3.
Parallel non-collocated sparse COLD array.

dipoles can be done to reduce the coupling effects even more [30]. This approach also allows for the independent reception of signal horizontal and vertical polarization components using orthogonal magnetic loops and dipoles.

Inspired by the design methods for parallel polarimetric array in [31, 32], [30] presents a design method for a parallel non-co-centered orthogonal loop and dipole (COLD) array as shown in **Figure 3**. The design concept of this array is as follows: (1) stretch dipole antennas and loop antennas in pairs along the z-axis; (2) ensure the distance between dipole antennas and loop antennas no greater than half a signal wavelength to avoid ambiguity in estimating 2D DOA; (3) configure dipole antennas and loop antennas at common sparse array positions (such as NAs/super nested arrays and ANAs).

2.2 Design of sparse polarimetric arrays based on uniform sparsity

The systematic design of element positions is required for non-uniform sparse polarimetric arrays, while the element positions of uniform sparse polarimetric arrays are uniformly sparsely distributed, making the optimization design process relatively simple. In practice, the "vector cross-product" characteristics of polarimetric arrays such as EMVA, triad-compositions of collocated dipoles/loops can be utilized to obtain uniform sparsity configuration of the array. Ref. [27] presents a sparse polarimetric array composed of EMVA elements arranged in a rectangular structure (as shown in **Figure 4**), which utilizes the array aperture expansion brought by sparsity and the "vector cross-product" characteristics of collocated EMVA to achieve higher precision in multi-dimensional parameter estimation.

Although the array shown in **Figure 4** can be used to improve the accuracy of multi-dimensional parameter estimation. However, similar to the sparse polarimetric array shown in **Figure 2**, the array shown in **Figure 4** also faces strong mutual coupling effects within the EMVA elements. A study in [29] proposes a spatially separated sparse rectangular polarimetric array composed of cross-dipoles, as shown in **Figure 5**. This array avoids potential mutual coupling errors by stretching all dipoles. It can achieve fast multi-dimensional parameter estimation with

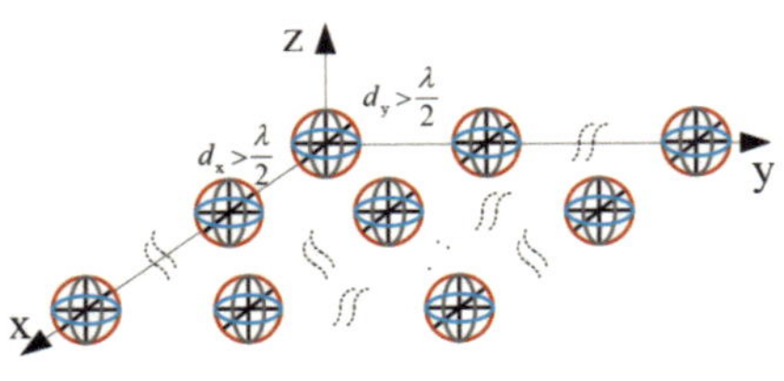

Figure 4.
Rectangular sparse polarimetric array constructed by EMVA.

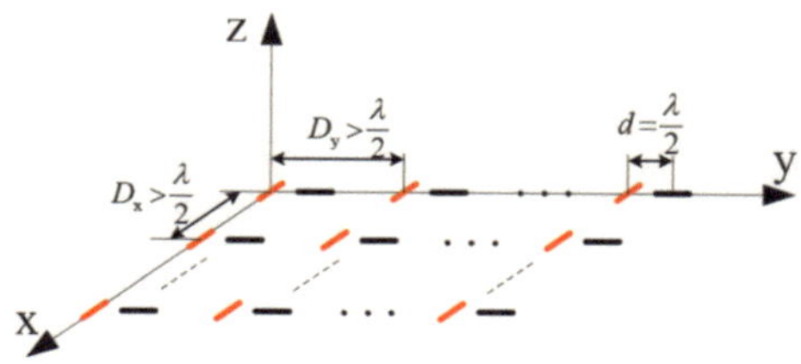

Figure 5.
Sparse rectangular polarimetric array composed of spatially distributed cross-dipoles.

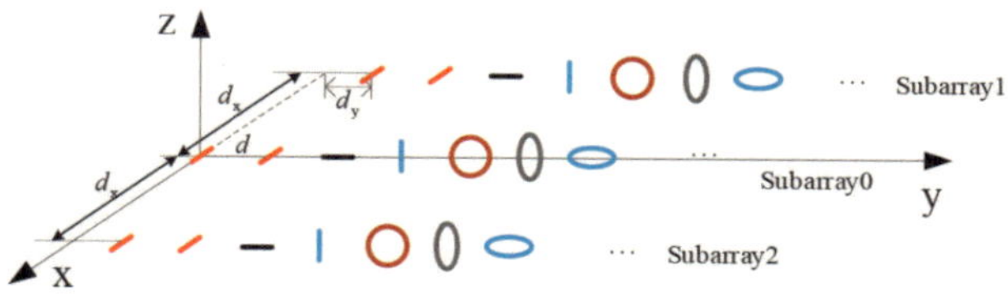

Figure 6.
Axial translation multilinear sparse polarimetric array.

high-precision based on its enlarged aperture array reconstruction sub-array translational invariance characteristics.

However, the above two arrays fail to effectively exploit the "vector cross-product" characteristics of polarimetric arrays such as EMVA and triad-compositions of collocated dipoles/loops. In order to fully utilize the potential of the "vector cross-product" characteristics in polarimetric arrays, [15] presents an axially parallel multilinear sparse polarimetric array as shown in **Figure 6**. It can fully utilize the "vector cross-product" characteristics and ultimately achieve high-precision unambiguous estimation of multiple parameters based on the virtual rotation invariance constructed among the sub-arrays.

2.3 Design of sparse polarimetric arrays based on mixed non-uniform and uniform sparsity

Instead of the design methods for uniform or non-uniform sparse polarimetric arrays, mixed uniform and non-uniform polarimetric arrays can enhance the diversity of polarimetric arrays and potentially improve the accuracy of parameter estimation. Presented below are two fundamental design methods for mixed uniform and non-uniform sparse polarimetric arrays:

2.3.1 Hybrid sparse polarimetric array with non-uniform sub-arrays uniformly spaced

Based on the spatially separated EMVA proposed in [28], **Figure 7** shows a hybrid sparse polarimetric array composed of spatially separated EMVA (non-uniform sparse) as sub-arrays uniformly spaced, where the internal structure of each basic unit and the final sparse array structure are shown. Ambiguity-free estimation can be achieved based on the translational invariance constructed by the uniform sparse array, and de-ambiguity of the ambiguous precise estimation can be achieved by utilizing the "vector cross-product" characteristics of spatially separated EMVA.

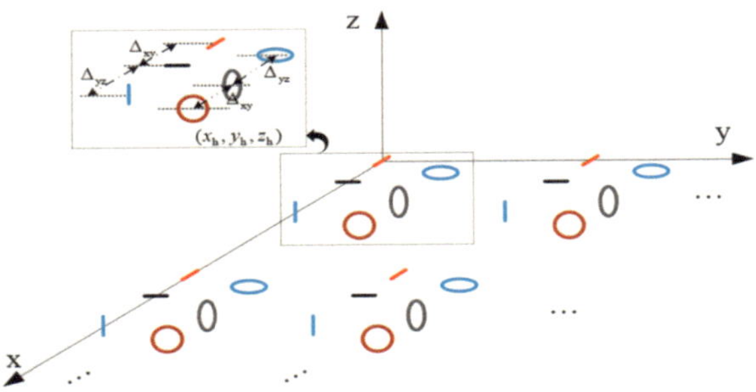

Figure 7.
Sparse polarimetric array composed of uniformly distributed non-uniform sparse sub-arrays.

2.3.2 Hybrid sparse polarimetric array with uniform sub-arrays non-uniformly spaced

Figure 8 shows a hybrid sparse polarimetric array under this scheme, which is mainly based on a uniform linear array with different polarizations parallel to the x-axis and non-uniform sparse placement along the y-axis to form a hybrid sparse polarimetric array [33]. The specific construction method is as follows: construct a cascaded sparse polarimetric planar array composed of $N_P \geq 2$ sparse sub-arrays cascaded together, with a distance d between adjacent sub-arrays. Each sub-array consists of $L_0 N_m$ magnetic loops or dipoles of the same polarization type, with the normal direction of the magnetic loop parallel to a certain coordinate axis and the axial direction of the dipole parallel to a certain coordinate axis. Each subarray is also composed of $N_m \geq 2$ mutually parallel layers, with a distance d_x between adjacent layers, and each layer has L_0 elements. Therefore, each layer of the cascaded sparse polarimetric planar array has a total of $L = L_0 N_P$ elements. The sparse arrangement of elements in each layer is a completely expandable sparse array. The number of virtual elements corresponding to each layer of each sub-array is L_V, with a spacing of d between adjacent virtual elements. It is worth noting that the design of sparse polarimetric arrays is a globally coordinated optimization process that needs to be systematically considered and evaluated from multiple perspectives, such as angular resolution, accuracy of multi-dimensional parameter estimation, beamforming performance, and computational complexity. The CSPA is easy to be manufactured and

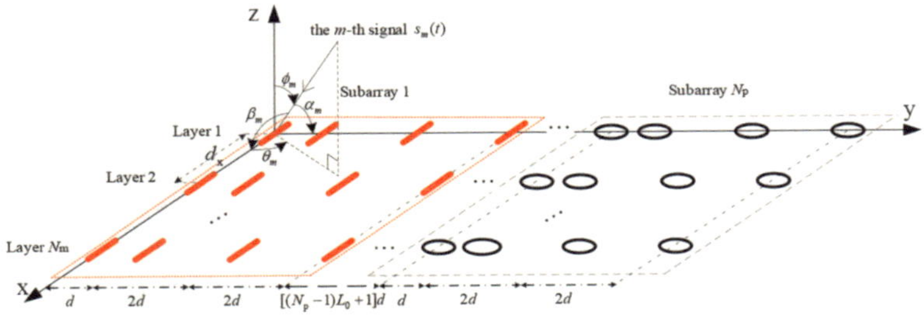

Figure 8.
One configuration example of the CSPA, where the antenna locations for each sparse linear array are given by (0, 1, 3, 5)*d. The red lines in Sub-array 1 represent the electrically* short *dipoles with axial directions parallel to the x-axis, and the gray ellipses in Sub-array* Np *represent the electrically* small *loops with normal directions parallel to the z-axis.*

less affected by mutual coupling that will lead to steering vector mismatch. It has enhanced DoFs, a closed-form expression for the antenna locations, and lower mutual coupling as compared with collocated diversely polarized antenna array.

3. Sparse polarimetric array signal model

Using the cascaded sparse polarimetric planar array (referred to as CSPA) depicted in **Figure 8** as a reference, the position matrices $\mathbf{V}_{n_p,n_m} \in \mathbb{R}^{L_0\times 2}$ of the L_0 elements in Sub-array n_p and Layer n_m in x-y Cartesian coordinates can be expressed as

$$\mathbf{V}_{n_p,n_m} = \left[(n_m-1)d_{\mathrm{x}}\mathbf{1}_{L_0}, (n_p-1)L_{\mathrm{v}}d + \mathbf{v}_{1,1}\right] \tag{1}$$

Here, $\mathbf{v}_{1,1}$ denotes the y-axis coordinates for the elements in Sub-array 1 and Layer 1. The sparse arrangement of the CSPA in each layer leads to an increased number of continuous lags in each layer, denoted as $N_{\mathrm{p}}(L_{\mathrm{v}}+1)-1$, with L_{v} being the number of virtual contiguous elements for each sub-array in one layer. The number of continuous lags in y-axis direction is N_{m}. Assuming one desired signal $s_0(t)$ and M interferences $\{s_m(t)\}_{m=1}^{M}$ imping on the CSPA, the array output is given by

$$\mathbf{x}(t) = \underbrace{\mathbf{a}_{\theta_0,\phi_0,\gamma_0,\eta_0}s_0(t)}_{=\mathbf{s}(t)} + \underbrace{\sum\nolimits_{m=1}^{M}\mathbf{a}_{\theta_m,\phi_m,\gamma_m,\eta_m}s_m(t)}_{=\mathbf{i}(t)} + \mathbf{n}(t), \tag{2}$$

in which $\mathbf{s}(t)$, $\mathbf{i}(t)$, and $\mathbf{n}(t)$ represent the independent array response components of the desired signal, interferences, and zero-mean spatially and temporally white Gaussian noise, respectively. $\mathbf{a}_{\theta_m,\phi_m,\gamma_m,\eta_m}$ represent the steering vectors of the m-th incident signal. The azimuth and elevation angles of the m-th signal are θ_m and ϕ_m, respectively. The angle between the m-th incident signal and the y-axis is denoted as α_m, while the angle between the m-th incident signal and the x-axis is denoted as β_m. The mathematical relationship between these four parameters can be written as

$$\sin\phi_m \sin\theta_m = \cos\alpha_m,\ \sin\phi_m\cos\theta_m = \cos\beta_m. \tag{3}$$

The steering vector of the m-th signal associated with the CSPA is

$$\mathbf{a}_{\theta_m,\phi_m,\gamma_m,\eta_m} = \mathrm{diag}(\mathbf{a}_{\mathrm{s},\,m})\mathbf{DJ\Xi}_{\theta_m,\phi_m}\mathbf{q}_{\gamma_m,\eta_m}, \tag{4}$$

where

$$\mathbf{a}_{\mathrm{s},\,m} = \left[\left(\mathbf{a}_{\mathrm{s},\,m}^{(1)}\right)^{\mathrm{T}}, \left(\mathbf{a}_{\mathrm{s},\,m}^{(2)}\right)^{\mathrm{T}}, \cdots, \left(\mathbf{a}_{\mathrm{s},\,m}^{(N_{\mathrm{m}})}\right)^{\mathrm{T}}\right]^{\mathrm{T}}. \tag{5}$$

The above equation is the spatial steering vector, in which $\mathbf{a}_{\mathrm{s},\,m}^{(1)}$, $\mathbf{a}_{\mathrm{s},\,m}^{(2)}$, and $\mathbf{a}_{\mathrm{s},\,m}^{(N_{\mathrm{m}})}$ are the spatial steering vectors of the m-th signal associated with the sub-array in Layer 1, Layer 2, and Layer N_{m} shown as follows:

$$\begin{aligned}\mathbf{a}_{\mathrm{s},\,m}^{(1)} &= \left[\mathrm{e}^{\mathrm{j}2\pi d_1\sin\phi_m\sin\theta_m/\lambda}, \mathrm{e}^{\mathrm{j}2\pi d_2\sin\phi_m\sin\theta_m/\lambda}, \cdots, \mathrm{e}^{\mathrm{j}2\pi d_L\sin\phi_m\sin\theta_m/\lambda}\right]^{\mathrm{T}},\\ \mathbf{a}_{\mathrm{s},\,m}^{(n_{\mathrm{m}})} &= \mathrm{e}^{\mathrm{j}2\pi(n_{\mathrm{m}}-1)d_{\mathrm{x}}\sin\phi_m\cos\theta_m/\lambda}\mathbf{a}_{\mathrm{s},\,m}^{(1)},\end{aligned} \tag{6}$$

with λ being the signal wavelength, d_l $(l = 1, 2, \cdots, L)$ being the distance between the l-th antenna and the coordinate origin, and $n_\mathrm{m} = 1, 2, \cdots, N_\mathrm{m}$. $\mathbf{D}\mathbf{J}\boldsymbol{\Xi}_{\theta_m,\phi_m}\mathbf{q}_{\gamma_m,\eta_m}$ is the polarimetric steering vector, where

$$\mathbf{D} = \left[\left(\mathbf{D}^{(1)}\right)^\mathrm{T}, \left(\mathbf{D}^{(2)}\right)^\mathrm{T}, \cdots, \left(\mathbf{D}^{(N_\mathrm{m})}\right)^\mathrm{T}\right]^\mathrm{T} \in \mathbb{R}^{N_\mathrm{m}L \times N_\mathrm{p}}, \tag{7}$$

is the directional weight component for the array elements $(\mathbf{D}^{(1)} = \mathbf{D}^{(2)} = \cdots = \mathbf{D}^{(N_\mathrm{m})})$ [33]. $\mathbf{J} \in \mathbb{N}^{N_\mathrm{p} \times 6}$ is the associated polarization selection matrix. $\boldsymbol{\Xi}_{\theta_m,\phi_m}\mathbf{q}_{\gamma_m,\eta_m}$ is the polarimetric steering vector of the electromagnetic vector antenna, and

$$\boldsymbol{\Xi}_{\theta_m,\phi_m} = \begin{bmatrix} -\sin\theta_m & \cos\phi_m\cos\theta_m \\ \cos\theta_m & \cos\phi_m\sin\theta_m \\ 0 & -\sin\phi_m \\ \cos\phi_m\cos\theta_m & \sin\theta_m \\ \cos\phi_m\sin\theta_m & -\cos\theta_m \\ -\sin\phi_m & 0 \end{bmatrix}, \tag{8}$$

$$\mathbf{q}_{\gamma_m,\eta_m} = \left[\cos\gamma_m, \sin\gamma_m \mathrm{e}^{\mathrm{j}\eta_m}\right]^\mathrm{T}, \tag{9}$$

in which γ_m and η_m are the auxiliary polarization angle and polarization phase difference of the m-th signal, respectively.

4. Closed-form multi-parameter estimation

In this section, we investigate a closed-form 2D DOA and polarization joint estimation method, known as decoupled sparse polarimetric rooting (DSPR), utilizing matrix transformation. The polynomial coefficients are derived for the general CSPA configuration with different pairs of N_m and N_p. Prior to delving into the detailed multi-parameter estimation procedures, we aim to construct a covariance matrix corresponding to the uniform counterpart of the array under consideration.

4.1 Multi-dimensional data fitting for the reconstructed covariance matrix

Based on the CSPA configuration, it can be concluded from (5) that

$$\mathbf{a}_{\mathrm{s},\,m}^{(n_\mathrm{m})} = \left[\left(\mathbf{a}_{\mathrm{s},\,m}^{(n_\mathrm{m},1)}\right)^\mathrm{T}, \left(\mathbf{a}_{\mathrm{s},\,m}^{(n_\mathrm{m},2)}\right)^\mathrm{T}, \cdots, \mathbf{a}_{\mathrm{s},\,m}^{(n_\mathrm{m},N_\mathrm{p})}\right]^\mathrm{T}. \tag{10}$$

The spatial steering vectors of different layers and different sub-arrays are given by

$$\mathbf{a}_{\mathrm{s},\,m}^{(n_\mathrm{m},n_\mathrm{p})} = \mathrm{e}^{\mathrm{j}2\pi(n_\mathrm{m}-1)d_\mathrm{x}\cos\beta_m/\lambda}\mathbf{a}_{\mathrm{s},\,m}^{(1,n_\mathrm{p})}, \tag{11}$$

with $n_\mathrm{p} = 1, 2, \cdots, N_\mathrm{p}$. According to (4), $\mathbf{a}_{\mathrm{s},\,m}$ can also be written as $\boldsymbol{\beta}_m \otimes \mathbf{a}_{\mathrm{s},\,m}^{(1)}$ with

$$\boldsymbol{\beta}_m = \left[1, e^{j2\pi d_x \cos\beta_m/\lambda}, \cdots, e^{j2\pi(N_m-1)d_x\cos\beta_m/\lambda}\right]^T \in \mathbb{C}^{N_m\times 1}. \quad (12)$$

Additionally, according to the mathematical structure of $\mathbf{DJ\Xi}_{\theta_m}\mathbf{q}_{\gamma_m,\eta_m}$ and the relationship between $\mathbf{DJ\Xi}_{\theta_m}\mathbf{q}_{\gamma_m,\eta_m}$ and $\mathbf{a}_{s,\,m}$, $\mathbf{a}_{\theta_m,\phi_m,\gamma_m,\eta_m}$ is also given by

$$\begin{aligned}\mathbf{a}_{\theta_m,\phi_m,\gamma_m,\eta_m} &= \text{diag}\left(\boldsymbol{\beta}_m \otimes \mathbf{a}^{(1)}_{s,\,m}\right)\mathbf{DJ\Xi}_{\theta_m,\phi_m}\mathbf{q}_{\gamma_m,\eta_m} = \left\{\mathbf{I}_{N_m}\otimes\left[\underbrace{\text{diag}\left(\mathbf{a}^{(1)}_{s,\,m}\right)\mathbf{D}^{(1)}\mathbf{J\Xi}_{\theta_m,\phi_m}\mathbf{q}_{\gamma_m,\eta_m}}_{=\mathbf{a}^{(1)}_{\theta_m,\phi_m,\gamma_m,\eta_m}}\right]\right\}\boldsymbol{\beta}_m \\ &= \left[\mathbf{I}_{N_m}\otimes \text{diag}\left(\mathbf{a}^{(1)}_{s,\,m}\right)\right]\left[\mathbf{I}_{N_m}\otimes\mathbf{D}^{(1)}\right]\left[\mathbf{I}_{N_m}\otimes\mathbf{J}\right]\left[\mathbf{I}_{N_m}\otimes\boldsymbol{\Xi}_{\theta_m,\phi_m}\right]\left[\mathbf{I}_{N_m}\otimes\mathbf{q}_{\gamma_m,\eta_m}\right]\boldsymbol{\beta}_m.\end{aligned} \quad (13)$$

Then, the array covariance matrix can be written as

$$\mathbf{R_{xx}} = \sum_{m=0}^{M}\sigma_m^2\begin{bmatrix} \mathbf{a}^{(1)}_{\theta_m,\phi_m,\gamma_m,\eta_m}\left[\mathbf{a}^{(1)}_{\theta_m,\phi_m,\gamma_m,\eta_m}\right]^H & \cdots & u_m^{1-N_m}\mathbf{a}^{(1)}_{\theta_m,\phi_m,\gamma_m,\eta_m}\left[\mathbf{a}^{(1)}_{\theta_m,\phi_m,\gamma_m,\eta_m}\right]^H \\ \vdots & \ddots & \vdots \\ u_m^{N_m-1}\mathbf{a}^{(1)}_{\theta_m,\phi_m,\gamma_m,\eta_m}\left[\mathbf{a}^{(1)}_{\theta_m,\phi_m,\gamma_m,\eta_m}\right]^H & \cdots & \mathbf{a}^{(1)}_{\theta_m,\phi_m,\gamma_m,\eta_m}\left[\mathbf{a}^{(1)}_{\theta_m,\phi_m,\gamma_m,\eta_m}\right]^H \end{bmatrix} + \sigma^2\mathbf{I}_{N_m L}, \quad (14)$$

where σ^2 is the power of the noise, $\sigma_m^2 (m = 0,1,2, \ldots, M)$ are the power of the desired signal or the M interferences, and $u_m = e^{j2\pi d_x\cos\beta_m/\lambda}$. The matrix $\mathbf{R_{xx}}$ can be partitioned into N_m^2 submatrices, and the array covariance matrix transformation could be decomposed into N_m^2 similar transformational procedures. Here, the transformation of the submatrix $\mathbf{R}^{(11)}_{\mathbf{xx}}$ (shown as (15)) in the left upper corner of $\mathbf{R_{xx}}$ is given as follows, and the transformations of the other $N_m^2 - 1$ submatrices are similar.

$$\mathbf{R}^{(11)}_{\mathbf{xx}} = \sum_{m=0}^{M}\sigma_m^2\begin{bmatrix} \delta_{11,m}\mathbf{a}^{(1,1)}_{s,m}\left[\mathbf{a}^{(1,1)}_{s,m}\right]^H & \cdots & \delta_{1N_p,m}\mathbf{a}^{(1,1)}_{s,m}\left[\mathbf{a}^{(1,N_p)}_{s,m}\right]^H \\ \vdots & \ddots & \vdots \\ \delta_{N_p1,m}\mathbf{a}^{(1,N_p)}_{s,m}\left[\mathbf{a}^{(1,1)}_{s,m}\right]^H & \cdots & \delta_{N_pN_p,m}\mathbf{a}^{(1,N_p)}_{s,m}\left[\mathbf{a}^{(1,N_p)}_{s,m}\right]^H \end{bmatrix} + \sigma^2\mathbf{I}_L, \quad (15)$$

in which $\delta_{n_1n_2,m} = [\boldsymbol{\delta}_m]_{n_1}[\boldsymbol{\delta}_m]^*_{n_2}$ and $\boldsymbol{\delta}_m = \mathbf{D}^{(1)}(1 : L_0 : \text{end}, :)\mathbf{J\Xi}_{\theta_m,\phi_m}\mathbf{q}_{\gamma_m,\eta_m}$ with $\mathbf{D}^{(1)}(1 : L_0 : \text{end}, :)$ denoting the submatrix composed of the first, $(1 + L_0) -$ th, $\cdots$, $[1 + (N_p - 1)L_0] -$ th rows of $\mathbf{D}^{(1)}$. It can then be learned from the data structure of $\mathbf{D}^{(1)}$ [33] that $\Sigma_{m=0}^{M}\sigma_m^2\mathbf{a}^{(1)}_{\theta_m,\phi_m,\gamma_m,\eta_m}\left[\mathbf{a}^{(1)}_{\theta_m,\phi_m,\gamma_m,\eta_m}\right]^H$ can be further simplified as

$$\sum_{m=0}^{M}\sigma_m^2\mathbf{a}^{(1)}_{\theta_m,\phi_m,\gamma_m,\eta_m}\left[\mathbf{a}^{(1)}_{\theta_m,\phi_m,\gamma_m,\eta_m}\right]^H = \sum_{m=0}^{M}\sigma_m^2\begin{bmatrix} \rho_{11,m}\mathbf{a}^{(1,1)}_{s,m}\left[\mathbf{a}^{(1,1)}_{s,m}\right]^H & \cdots & \rho_{1N_p,m}\mathbf{a}^{(1,1)}_{s,m}\left[\mathbf{a}^{(1,1)}_{s,m}\right]^H \\ \vdots & \ddots & \vdots \\ \rho_{N_p1,m}\mathbf{a}^{(1,1)}_{s,m}\left[\mathbf{a}^{(1,1)}_{s,m}\right]^H & \cdots & \rho_{N_pN_p,m}\mathbf{a}^{(1,1)}_{s,m}\left[\mathbf{a}^{(1,1)}_{s,m}\right]^H \end{bmatrix}, \quad (16)$$

with $\rho_{11,m} = \delta_{11,m}, \rho_{12,m} = \mathrm{e}^{-\mathrm{j}2\pi L_{\mathrm{v}} d \cos\alpha_m/\lambda}\delta_{12,m}, \cdots, \rho_{1N_{\mathrm{p}},m} = \mathrm{e}^{-\mathrm{j}2\pi\left(N_{\mathrm{p}}-1\right)L_{\mathrm{v}} d \cos\alpha_m/\lambda}$ $\delta_{1N_{\mathrm{p}},m}, \rho_{21,m} = \mathrm{e}^{\mathrm{j}2\pi L_{\mathrm{v}} d \cos\alpha_m/\lambda}\delta_{21,m}, \rho_{22,m} = \delta_{12,m}, \cdots, \rho_{2N_{\mathrm{p}},m} = \mathrm{e}^{-\mathrm{j}2\pi\left(N_{\mathrm{p}}-2\right)L_{\mathrm{v}} d \cos\alpha_m/\lambda}$ $\delta_{2N_{\mathrm{p}},m}, \cdots, \rho_{N_{\mathrm{p}}1,m} = \mathrm{e}^{\mathrm{j}2\pi\left(N_{\mathrm{p}}-1\right)L_{\mathrm{v}} d \cos\alpha/\lambda}\delta_{N_{\mathrm{p}}1,m}, \rho_{N_{\mathrm{p}}2,m} = \mathrm{e}^{\mathrm{j}2\pi\left(N_{\mathrm{p}}-2\right)L_{\mathrm{v}} d \cos\alpha/\lambda}$ $\delta_{N_{\mathrm{p}}2,m}, \cdots, \rho_{N_{\mathrm{p}}N_{\mathrm{p}},m} = \delta_{N_{\mathrm{p}}N_{\mathrm{p}},m}$. Therefore, $\mathbf{R_{xx}}$ can be divided into $\left(N_{\mathrm{m}}N_{\mathrm{p}}\right)^2$ submatrices, and all transformations of these submatrices are similar. Here, we provide the transformation of the submatrix $\mathbf{R}_{11}^{(11)} = \mathbf{R}_{\mathbf{xx}}^{(11)}(1:L_0, 1:L_0)$ as an illustration. By vectorizing $\mathbf{R}_{11}^{(11)}$, we can derive the submatrix corresponding to a virtual uniform linear array consisting of L_{v} elements with the same polarization type as the antennas in the first sub-array, i.e.,

$$\tilde{\mathbf{R}}_{11}^{(11)} = \sum_{m=0}^{M} \rho_{11,m}\sigma_m^2 \tilde{\mathbf{a}}_{\mathrm{s},m}^{(1,1)}\left[\tilde{\mathbf{a}}_{\mathrm{s},m}^{(1,1)}\right]^{\mathrm{H}} + \sigma^2\mathbf{I}_{L_{\mathrm{v}}}, \tag{17}$$

with $\tilde{\mathbf{a}}_{\mathrm{s},m}^{(1,1)} = \left[1, \mathrm{e}^{\mathrm{j}2\pi d\cos\alpha_m/\lambda}, \cdots, \mathrm{e}^{\mathrm{j}2\pi L_{\mathrm{v}} d\cos\alpha_m/\lambda}\right]^{\mathrm{T}} \in \mathbb{C}^{L_{\mathrm{v}}\times 1}$. Finally, by applying a similar transformation, the matrix $\mathbf{R}_{\mathbf{xx}}^{(11)}$ can be converted into $\mathbf{R}_{\tilde{\mathbf{x}}\tilde{\mathbf{x}}}^{(11)}$ shown below:

$$\begin{aligned}\mathbf{R}_{\tilde{\mathbf{x}}\tilde{\mathbf{x}}}^{(11)} &= \sum_{m=0}^{M}\sigma_m^2\tilde{\mathbf{a}}_{\theta_m,\phi_m,\gamma_m,\eta_m}^{(1)}\left[\tilde{\mathbf{a}}_{\theta_m,\phi_m,\gamma_m,\eta_m}^{(1)}\right]^{\mathrm{H}} + \sigma^2\mathbf{I}_{L_{\mathrm{v}}N_{\mathrm{p}}} = \sigma^2\mathbf{I}_{L_{\mathrm{v}}N_{\mathrm{p}}} \\ &+ \sum_{m=0}^{M}\sigma_m^2\begin{bmatrix}\rho_{11,m}\tilde{\mathbf{a}}_{\mathrm{s},m}^{(1,1)}\left[\tilde{\mathbf{a}}_{\mathrm{s},m}^{(1,1)}\right]^{\mathrm{H}} & \cdots & \rho_{1N_{\mathrm{p}},m}\tilde{\mathbf{a}}_{\mathrm{s},m}^{(1,1)}\left[\tilde{\mathbf{a}}_{\mathrm{s},m}^{(1,1)}\right]^{\mathrm{H}} \\ \vdots & \ddots & \vdots \\ \rho_{N_{\mathrm{p}}1,m}\tilde{\mathbf{a}}_{\mathrm{s},m}^{(1,1)}\left[\tilde{\mathbf{a}}_{\mathrm{s},m}^{(1,1)}\right]^{\mathrm{H}} & \cdots & \rho_{N_{\mathrm{p}}N_{\mathrm{p}},m}\tilde{\mathbf{a}}_{\mathrm{s},m}^{(1,1)}\left[\tilde{\mathbf{a}}_{\mathrm{s},m}^{(1,1)}\right]^{\mathrm{H}}\end{bmatrix}.\end{aligned} \tag{18}$$

where $\Sigma_{m=0}^{M}\sigma_m^2\tilde{\mathbf{a}}_{\theta_m,\phi_m,\gamma_m,\eta_m}\tilde{\mathbf{a}}_{\theta_m,\phi_m,\gamma_m,\eta_m}^{\mathrm{H}}$ is also given by

$$\sum_{m=0}^{M}\sigma_m^2\begin{bmatrix}\delta_{11,m}\tilde{\mathbf{a}}_{\mathrm{s},m}^{(1,1)}\left[\tilde{\mathbf{a}}_{\mathrm{s},m}^{(1,1)}\right]^{\mathrm{H}} & \cdots & \delta_{1N_{\mathrm{p}},m}\tilde{\mathbf{a}}_{\mathrm{s},m}^{(1,1)}\left[\tilde{\mathbf{a}}_{\mathrm{s},m}^{(1,N_{\mathrm{p}})}\right]^{\mathrm{H}} \\ \vdots & \ddots & \vdots \\ \delta_{N_{\mathrm{p}}1,m}\tilde{\mathbf{a}}_{\mathrm{s},m}^{(1,N_{\mathrm{p}})}\left[\tilde{\mathbf{a}}_{\mathrm{s},m}^{(1,1)}\right]^{\mathrm{H}} & \cdots & \delta_{N_{\mathrm{p}}N_{\mathrm{p}},m}\tilde{\mathbf{a}}_{\mathrm{s},m}^{(1,N_{\mathrm{p}})}\left[\tilde{\mathbf{a}}_{\mathrm{s},m}^{(1,N_{\mathrm{p}})}\right]^{\mathrm{H}}\end{bmatrix}, \tag{19}$$

The transformation of the covariance matrix entails several steps, including partitioning, vectorizing, and recombining operations [33, 34] on these $\left(N_{\mathrm{m}}N_{\mathrm{p}}\right)^2$ submatrices in $\mathbf{R_{xx}}$. By employing similar partitioning, vectorizing, dividing, and recombining operations on these $\left(N_{\mathrm{m}}N_{\mathrm{p}}\right)^2$ submatrices in $\mathbf{R_{xx}}$, a higher-dimensional covariance matrix $\mathbf{R}_{\tilde{\mathbf{x}}\tilde{\mathbf{x}}}$ is reconstructed. In essence, through these above-mentioned operations, the covariance matrix $\mathbf{R}_{\tilde{\mathbf{x}}\tilde{\mathbf{x}}}$ can be derived from the initial covariance matrix $\mathbf{R_{xx}}$, and

$$\mathbf{R}_{\tilde{\mathbf{x}}\tilde{\mathbf{x}}} = \sum_{m=0}^{M}\sigma_m^2\tilde{\mathbf{a}}_{\theta_m,\phi_m,\gamma_m,\eta_m}\tilde{\mathbf{a}}_{\theta_m,\phi_m,\gamma_m,\eta_m}^{\mathrm{H}} + \sigma^2\mathbf{I}_{L_{\mathrm{v}}N_{\mathrm{m}}N_{\mathrm{p}}} \tag{20}$$

is linked to the virtual cascaded uniform planar array (CUPA) comprising antennas with diverse polarizations, where the steering vector $\tilde{\mathbf{a}}_{\theta_m,\phi_m,\gamma_m,\eta_m}$ can be expressed as:

$$\tilde{\mathbf{a}}_{\theta_m,\phi_m,\gamma_m,\eta_m} = \text{diag}\left(\boldsymbol{\beta}_m \otimes \tilde{\mathbf{a}}_{\text{s},m}^{(1)}\right)\tilde{\mathbf{D}}\mathbf{J}\boldsymbol{\Xi}_{\theta_m,\phi_m}\mathbf{q}_{\gamma_m,\eta_m} = \left\{\mathbf{I}_{N_\text{m}} \otimes \left[\text{diag}\left(\tilde{\mathbf{a}}_{\text{s},m}^{(1)}\right)\tilde{\mathbf{D}}^{(1)}\mathbf{J}\boldsymbol{\Xi}_{\theta_m,\phi_m}\mathbf{q}_{\gamma_m,\eta_m}\right]\right\}\boldsymbol{\beta}_m$$
$$= \left[\mathbf{I}_{N_\text{m}} \otimes \text{diag}\left(\tilde{\mathbf{a}}_{\text{s},m}^{(1)}\right)\right]\left[\mathbf{I}_{N_\text{m}} \otimes \tilde{\mathbf{D}}^{(1)}\right]\left[\mathbf{I}_{N_\text{m}} \otimes \mathbf{J}\right]\left[\mathbf{I}_{N_\text{m}} \otimes \boldsymbol{\Xi}_{\theta_m,\phi_m}\right]\left[\mathbf{I}_{N_\text{m}} \otimes \mathbf{q}_{\gamma_m,\eta_m}\right]\boldsymbol{\beta}_m, \tag{21}$$

with $\tilde{\mathbf{a}}_{\text{s},m}^{(1)} = \left[\left(\tilde{\mathbf{a}}_{\text{s},m}^{(1,1)}\right)^\text{T}, \left(\tilde{\mathbf{a}}_{\text{s},m}^{(1,2)}\right)^\text{T}, \cdots, \tilde{\mathbf{a}}_{\text{s},m}^{(1,N_\text{p})}\right]^\text{T} \in \mathbb{C}^{N_\text{p}L_\text{v}\times 1}$, in which

$$\tilde{\mathbf{a}}_{\text{s},m}^{(1,n_\text{p})} = \text{e}^{\text{j}2\pi(n_\text{p}-1)L_\text{v}d\cos\alpha_m/\lambda}\tilde{\mathbf{a}}_{\text{s},m}^{(1,1)} = \left[\text{e}^{\text{j}2\pi(n_\text{p}-1)L_\text{v}d\cos\alpha_m/\lambda}, \cdots, \text{e}^{\text{j}2\pi(n_\text{p}L_\text{v}-1)d\cos\alpha_m/\lambda}\right]^\text{T}, \tag{22}$$

and $\tilde{\mathbf{D}} = \mathbf{1}_{N_\text{m}} \otimes \tilde{\mathbf{D}}^{(1)}$, where $\tilde{\mathbf{D}}^{(1)} \in \mathbb{R}^{(N_\text{p}L_\text{v})\times N_\text{p}}$ is the directional weight component linked to Layer 1 of the virtual CUPA. The antenna polarization in each sub-array of the virtual CUPA follows the same manner as that of the physical CSPA, and all the data of the $(N_\text{m}N_\text{p})^2$ submatrices in $\mathbf{R}_{\mathbf{xx}}$ are utilized in the data reconstruction process.

4.2 Multi-dimensional parameter estimation approach in closed form

Drawing from the principles of polarimetric multiple signal classification (MUSIC) [35], we can infer that

$$\boldsymbol{\beta}_m^\text{H}\left[\mathbf{I}_{N_\text{m}} \otimes \mathbf{q}_{\gamma_m,\eta_m}\right]^\text{H}\left[\mathbf{I}_{N_\text{m}} \otimes \boldsymbol{\Xi}_{\theta_m,\phi_m}\right]^\text{H}\left[\mathbf{I}_{N_\text{m}} \otimes \mathbf{J}\right]^\text{H}\left[\mathbf{I}_{N_\text{m}} \otimes \tilde{\mathbf{D}}^{(1)}\right]^\text{H}\left[\mathbf{I}_{N_\text{m}} \otimes \text{diag}\left(\tilde{\mathbf{a}}_{\text{s},m}^{(1)}\right)\right]^\text{H}\tilde{\mathbf{U}}\tilde{\mathbf{U}}^\text{H}$$
$$\left[\mathbf{I}_{N_\text{m}} \otimes \text{diag}\left(\tilde{\mathbf{a}}_{\text{s},m}^{(1)}\right)\right]\left[\mathbf{I}_{N_\text{m}} \otimes \tilde{\mathbf{D}}^{(1)}\right]\left[\mathbf{I}_{N_\text{m}} \otimes \mathbf{J}\right]\left[\mathbf{I}_{N_\text{m}} \otimes \boldsymbol{\Xi}_{\theta_m,\phi_m}\right]\left[\mathbf{I}_{N_\text{m}} \otimes \mathbf{q}_{\gamma_m,\eta_m}\right]\boldsymbol{\beta}_m = 0, \tag{23}$$

where $\tilde{\mathbf{U}}$ is the matrix whose columns correspond to the $L_\text{v}N_\text{m}N_\text{p} - M - 1$ smaller eigenvalues of $\mathbf{R}_{\tilde{\mathbf{x}}\tilde{\mathbf{x}}}$. Referring to (23) and the rank reduction principle, we obtain

$$\det\left\{\left[\mathbf{I}_{N_\text{m}} \otimes \tilde{\mathbf{D}}^{(1)}\right]^\text{H}\left[\mathbf{I}_{N_\text{m}} \otimes \text{diag}\left(\tilde{\mathbf{a}}_{\text{s},m}^{(1)}\right)\right]^\text{H}\tilde{\mathbf{U}}\tilde{\mathbf{U}}^\text{H}\left[\mathbf{I}_{N_\text{m}} \otimes \text{diag}\left(\tilde{\mathbf{a}}_{\text{s},m}^{(1)}\right)\right]\left[\mathbf{I}_{N_\text{m}} \otimes \tilde{\mathbf{D}}^{(1)}\right]\right\} = 0, \tag{24}$$

which means that $\left\{\text{e}^{\text{j}2\pi d\cos\alpha_m/\lambda}\right\}_{m=0}^{M}$ are the roots of the following equation

$$\det\left\{\left[\mathbf{I}_{N_\text{m}} \otimes \tilde{\mathbf{D}}^{(1)}\right]^\text{H}\boldsymbol{\Lambda}\left(z^{-1}\right)\mathbf{U}\mathbf{U}^\text{H}\boldsymbol{\Lambda}(z)\left[\mathbf{I}_{N_\text{m}} \otimes \tilde{\mathbf{D}}^{(1)}\right]\right\} = 0, \tag{25}$$

with $\boldsymbol{\Lambda}(z) = \mathbf{I}_{N_\text{m}} \otimes \text{diag}\left\{\left[1, z, \cdots, z^{L_\text{v}N_\text{p}-1}\right]^\text{T}\right\}$. Therefore, the polynomial rooting-based procedure would be employed to estimate α_m, and the polynomial coefficients can be computed using the matrix determinant calculation principle. For mathematical simplicity, $\tilde{\mathbf{D}}^{(1)}$ can be assumed to be $\mathbf{I}_{N_\text{p}} \otimes \mathbf{1}_{L_\text{v}}$ without loss of generality. $\boldsymbol{\Lambda}(z)\mathbf{D}$ in (25) can be reformulated as

$$\boldsymbol{\Lambda}(z)\mathbf{D} = \mathbf{I}_{N_\text{m}} \otimes \text{blkdiag}\left[\mathbf{h}_1(z), \mathbf{h}_2(z), \cdots, \mathbf{h}_{N_\text{p}}(z)\right] \tag{26}$$

in which

$$\mathbf{h}_{n_\mathrm{p}}(z) = \left[z^{(n_\mathrm{p}-1)L_\mathrm{v}}, z^{(n_\mathrm{p}-1)L_\mathrm{v}+1}, \ldots, z^{(n_\mathrm{p}-1)L_\mathrm{v}-1}\right]^\mathrm{T} \tag{27}$$

Then, according to (25), the following equation holds:

$$\det\left\{\mathbf{D}^\mathrm{H}\boldsymbol{\Lambda}(z^{-1})\underbrace{\begin{bmatrix} \mathbf{P}_{1,1} & \mathbf{P}_{1,2} & \cdots & \mathbf{P}_{1,N_\mathrm{m}N_\mathrm{p}} \\ \mathbf{P}_{2,1} & \mathbf{P}_{2,2} & \cdots & \mathbf{P}_{2,N_\mathrm{m}N_\mathrm{p}} \\ \vdots & \vdots & \ddots & \vdots \\ \mathbf{P}_{N_\mathrm{m}N_\mathrm{p},1} & \mathbf{P}_{N_\mathrm{m}N_\mathrm{p},2} & \cdots & \mathbf{P}_{N_\mathrm{m}N_\mathrm{p},N_\mathrm{m}N_\mathrm{p}} \end{bmatrix}}_{=\mathbf{P}}\boldsymbol{\Lambda}(z)\mathbf{D}\right\} = 0, \tag{28}$$

where $\mathbf{P} = \mathbf{U}\mathbf{U}^\mathrm{H} \in \mathbb{C}^{(L_\mathrm{v}N_\mathrm{m}N_\mathrm{p})\times(L_\mathrm{v}N_\mathrm{m}N_\mathrm{p})}$ is divided into $(N_\mathrm{m}N_\mathrm{p})^2$ submatrices of the same dimension as illustrated in (28). From this, we can deduce that:

$$\det\left\{\begin{bmatrix} \mathbf{h}_1^\mathrm{T}(z^{-1})\mathbf{P}_{11}\mathbf{h}_1(z) & \mathbf{h}_1^\mathrm{T}(z^{-1})\mathbf{P}_{12}\mathbf{h}_2(z) & \cdots & \mathbf{h}_1^\mathrm{T}(z^{-1})\mathbf{P}_{1N_\mathrm{p}}\mathbf{h}_{N_\mathrm{p}}(z) & \cdots & \mathbf{h}_1^\mathrm{T}(z^{-1})\mathbf{P}_{1(N_\mathrm{p}N_\mathrm{m})}\mathbf{h}_{N_\mathrm{p}}(z) \\ \mathbf{h}_2^\mathrm{T}(z^{-1})\mathbf{P}_{21}\mathbf{h}_1(z) & \mathbf{h}_2^\mathrm{T}(z^{-1})\mathbf{P}_{22}\mathbf{h}_2(z) & \cdots & \mathbf{h}_2^\mathrm{T}(z^{-1})\mathbf{P}_{2N_\mathrm{p}}\mathbf{h}_{N_\mathrm{p}}(z) & \cdots & \mathbf{h}_2^\mathrm{T}(z^{-1})\mathbf{P}_{2(N_\mathrm{p}N_\mathrm{m})}\mathbf{h}_{N_\mathrm{p}}(z) \\ \vdots & \vdots & \ddots & \vdots & \ddots & \vdots \\ \mathbf{h}_{N_\mathrm{p}}^\mathrm{T}(z^{-1})\mathbf{P}_{N_\mathrm{p}1}\mathbf{h}_1(z) & \mathbf{h}_{N_\mathrm{p}}^\mathrm{T}(z^{-1})\mathbf{P}_{N_\mathrm{p}2}\mathbf{h}_2(z) & \cdots & \mathbf{h}_{N_\mathrm{p}}^\mathrm{T}(z^{-1})\mathbf{P}_{N_\mathrm{p}N_\mathrm{p}}\mathbf{h}_{N_\mathrm{p}}(z) & \cdots & \mathbf{h}_{N_\mathrm{p}}^\mathrm{T}(z^{-1})\mathbf{P}_{N_\mathrm{p}(N_\mathrm{p}N_\mathrm{m})}\mathbf{h}_{N_\mathrm{p}}(z) \\ \vdots & \vdots & \ddots & \vdots & \ddots & \vdots \\ \mathbf{h}_{N_\mathrm{p}}^\mathrm{T}(z^{-1})\mathbf{P}_{(N_\mathrm{p}N_\mathrm{m})1}\mathbf{h}_1(z) & \mathbf{h}_{N_\mathrm{p}}^\mathrm{T}(z^{-1})\mathbf{P}_{(N_\mathrm{p}N_\mathrm{m})2}\mathbf{h}_2(z) & \cdots & \mathbf{h}_{N_\mathrm{p}}^\mathrm{T}(z^{-1})\mathbf{P}_{(N_\mathrm{p}N_\mathrm{m})N_\mathrm{p}}\mathbf{h}_{N_\mathrm{p}}(z) & \cdots & \mathbf{h}_{N_\mathrm{p}}^\mathrm{T}(z^{-1})\mathbf{P}_{(N_\mathrm{p}N_\mathrm{m})(N_\mathrm{p}N_\mathrm{m})}\mathbf{h}_{N_\mathrm{p}}(z) \end{bmatrix}\right\} = 0. \tag{29}$$

where the elements $\mathbf{h}_{n_\mathrm{p}}^\mathrm{T}(z^{-1})\mathbf{P}_{(n_\mathrm{p}n_\mathrm{m})(n_\mathrm{p}n_\mathrm{m})}\mathbf{h}_{n_\mathrm{p}}(z)$ in (29) can be obtained, and the final polynomial coefficients can be derived using the Laplace theorem. Simplified derivations are available in Refs. [36, 37]. It is evident that the polynomial rooting-based closed-form approach entails two crucial processes: the transformation of the array covariance matrix (involving partitioning, vectorizing, dividing, and recombining operations) and the calculation of polynomial coefficients. The angles α_m of the desired signal and interferences can be estimated by

$$\alpha_m = \arccos[\lambda(\angle z_m)/2\pi d], \tag{30}$$

Here, the roots of (25) are denoted by $\{z_m\}_{m=0}^{M}$ whose moduli lie inside and closest to the unit circle. According to (13) and rank reduction principle, the eigenvector corresponding to the smallest eigenvalue of $\left[\mathbf{I}_{N_\mathrm{m}} \otimes \mathbf{D}^{(1)}\right]^\mathrm{H}\left[\mathbf{I}_{N_\mathrm{m}} \otimes \mathrm{diag}\left(\mathbf{a}_{\mathrm{s},m}^{(1)}\right)\right]^\mathrm{H}\mathbf{U}\mathbf{U}^\mathrm{H}$ $\left[\mathbf{I}_{N_\mathrm{m}} \otimes \mathrm{diag}\left(\mathbf{a}_{\mathrm{s},m}^{(1)}\right)\right]\left[\mathbf{I}_{N_\mathrm{m}} \otimes \mathbf{D}^{(1)}\right]$ ($\mathbf{U}$ is the matrix comprising the eigenvectors linked with the $N_\mathrm{m}L - M - 1$ smaller eigenvalues of $\mathbf{R}_{\mathbf{xx}}$) can be expressed as

$$\mathbf{h}_m = h_m[\mathbf{I}_{N_\mathrm{m}} \otimes \mathbf{J}]\left[\mathbf{I}_{N_\mathrm{m}} \otimes \boldsymbol{\Xi}_{\theta_m,\phi_m}\right]\left[\mathbf{I}_{N_\mathrm{m}} \otimes \mathbf{q}_{\gamma_m,\eta_m}\right]\boldsymbol{\beta}_m = h_m\begin{bmatrix} \mathbf{J}\boldsymbol{\Xi}_{\theta_m,\phi_m}\mathbf{q}_{\gamma_m,\eta_m} \\ \mathrm{e}^{\mathrm{j}2\pi d_\mathrm{x}\cos\beta_m/\lambda}\mathbf{J}\boldsymbol{\Xi}_{\theta_m,\phi_m}\mathbf{q}_{\gamma_m,\eta_m} \\ \vdots \\ \mathrm{e}^{\mathrm{j}2\pi(N_\mathrm{m}-1)d_\mathrm{x}\cos\beta/\lambda}\mathbf{J}\boldsymbol{\Xi}_{\theta_m,\phi_m}\mathbf{q}_{\gamma_m,\eta_m} \end{bmatrix}, \tag{31}$$

in which h_m represents a non-zero constant. The subsequent equation is established by leveraging all the information contained within the vector $\mathbf{h}_m$:

$$\beta_m = \arccos\left\{\frac{\lambda}{2\pi d_\mathrm{x}}\angle\left[\sum_{n_1=1}^{N_\mathrm{m}-1}\sum_{n_2=1}^{N_\mathrm{p}}\frac{[\mathbf{h}_m]_{n_1 N_\mathrm{p}+n_2}}{[\mathbf{h}_m]_{(n_1-1)N_\mathrm{p}+n_2}}\right]\right\}, \tag{32}$$

The 2D DOAs of the desired signal and interferences, denoted by $\{\theta_m, \phi_m\}_{m=0}^{M}$, can be estimated using (3) without pair matching. Moreover, it is evident that the eigenvector corresponding to the smallest eigenvalue of

$$\begin{aligned}&\left[\mathbf{I}_{N_\mathrm{m}}\otimes\mathbf{\Xi}_{\theta_m,\phi_m}\right]^\mathrm{H}\left[\mathbf{I}_{N_\mathrm{m}}\otimes\mathbf{J}\right]^\mathrm{H}\left[\mathbf{I}_{N_\mathrm{m}}\otimes\mathbf{D}^{(1)}\right]^\mathrm{H}\left[\mathbf{I}_{N_\mathrm{m}}\otimes\mathrm{diag}\left(\mathbf{a}_{\mathrm{s},m}^{(1)}\right)\right]^\mathrm{H}\\&\mathbf{U}\mathbf{U}^\mathrm{H}\left[\mathbf{I}_{N_\mathrm{m}}\otimes\mathrm{diag}\left(\mathbf{a}_{\mathrm{s},m}^{(1)}\right)\right]\left[\mathbf{I}_{N_\mathrm{m}}\otimes\mathbf{D}^{(1)}\right]\left[\mathbf{I}_{N_\mathrm{m}}\otimes\mathbf{J}\right]\left[\mathbf{I}_{N_\mathrm{m}}\otimes\mathbf{\Xi}_{\theta_m,\phi_m}\right]\end{aligned} \tag{33}$$

can be expressed as

$$\mathbf{v}_m = v_m\left[\mathbf{I}_{N_\mathrm{m}}\otimes\mathbf{q}_{\gamma_m,\eta_m}\right]\boldsymbol{\beta}_m = v_m\left[\mathbf{q}_{\gamma_m,\eta_m}^\mathrm{T},\quad \mathrm{e}^{\mathrm{j}2\pi d_\mathrm{x}\cos\beta_m/\lambda}\mathbf{q}_{\gamma_m,\eta_m}^\mathrm{T},\quad \cdots,\quad \mathrm{e}^{\mathrm{j}2\pi(N_\mathrm{m}-1)d_\mathrm{x}\cos\beta_m/\lambda}\mathbf{q}_{\gamma_m,\eta_m}^\mathrm{T}\right] \tag{34}$$

with v_m being a non-zero constant. Consequently, one can ascertain the polarization parameters of both the desired signal and interferences via

$$\gamma_m = \arctan\left\{\frac{1}{N_\mathrm{m}}\left|\sum_{n=0}^{N_\mathrm{m}-1}\frac{[\mathbf{v}_m]_{2n+2}}{[\mathbf{v}_m]_{2n+1}}\right|\right\}, \tag{35}$$

$$\eta_m = \angle\left\{\sum_{n=0}^{N_\mathrm{m}-1}\frac{[\mathbf{v}_m]_{2n+2}}{[\mathbf{v}_m]_{2n+1}}\right\}. \tag{36}$$

Thus, the estimated steering vector $\mathbf{a}_{\hat{\theta}_0,\hat{\phi}_0,\hat{\gamma}_0,\hat{\eta}_0}$ in (39) can be derived. The DSPR method divides the 4D parameter estimation problem into sequential 1D parameter estimation processes (as described in (24) and (31), respectively), followed by a subsequent 2D parameter estimation procedure. This approach allows for efficient estimation of multi-dimensional parameters, streamlining the process and saving time.

5. Beamforming in joint spatial and polarimetric domain via sparse reconstruction

Adaptive beamforming is widely applied in radar, sonar, communications, acoustics, and medical imaging [38–41]. Numerous robust adaptive beamformers have been developed to enhance signal quality and mitigate interferences, addressing challenges such as model mismatches. Examples include the diagonal loading sample matrix inversion (DLSMI) beamformer, worst-case beamformer, Eigenspace beamformer, minimum variance distortionless response (MVDR) beam-former, unit circle roots constraints MVDR [42], interference covariance matrix reconstruction beamformer [43, 44], and interference covariance matrix sparse reconstruction beamformer [45, 46].

While traditional scalar arrays lack the ability to handle the polarization diversity of source signals, polarimetric arrays offer the advantage of leveraging signal polarization differences [2–13]. In this context, a polarimetric beamformer known as the polarimetric sparse-reconstruction (PSR) beamformer is introduced for the CSPA, jointly addressing spatial and polarimetric parameters from a sparse reconstruction perspective. This capability is crucial in complex interference scenarios [47, 48], such as when interferences coincide with the main-lobe [49–51] or when spatial separation between the desired signal and interferences is challenging. Designing adaptive beamformers with lower computational complexity is a significant challenge due to intensive parameter calculations. The PSR beamformer, developed within a reconstruction-based framework, exploits the sparsity of both desired signals and interferences in the joint spatial and polarimetric domain. Unlike beamformers for arrays with different polarization characteristics, which require high-dimensional spectral integrals, the PSR beamformer eliminates the need for such intensive calculations, ensuring computational efficiency.

In the realm of reconstruction-based beamformer, the computationally efficient PSR beamformer is introduced for the CSPA. It leverages the sparsity of both desired signals and interferences in the joint spatial and polarimetric domain. The computational efficiency of the PSR beamformer sets it apart from beamformers designed for single, dual, or diversely-polarized arrays, which typically involve computationally intensive 1D or 4D spectral integrals (e.g., Capon [52] or MUSIC methods) [43, 47]. Unlike the Integral-4D beamformer [47], which requires integral operations, the PSR beamformer avoids such complexities, offering a computationally attractive solution.

Let $\mathbf{w}$ denote the beamforming vector, the array output signal-to-interference-plus-noise ratio (SINR) is

$$\mathrm{SINR} = \frac{\sigma_0^2 \left|\mathbf{w}^{\mathrm{H}} \mathbf{a}_{\theta_0,\phi_0,\gamma_0,\eta_0}\right|^2}{\mathbf{w}^{\mathrm{H}} \mathbf{R}_{\mathbf{i+n}} \mathbf{w}}, \tag{37}$$

where σ_0^2 is the power of the desired signal, and $\mathbf{R}_{\mathbf{i+n}} = E\left\{[\mathbf{i}(t) + \mathbf{n}(t)][\mathbf{i}(t) + \mathbf{n}(t)]^{\mathrm{H}}\right\}$ is the interference-plus-noise covariance matrix. Maximizing SINR in (37) is equivalently represented by

$$\min_{\mathbf{w}} \mathbf{w}^{\mathrm{H}} \mathbf{R}_{\mathbf{i+n}} \mathbf{w} \qquad \text{subject to } \mathbf{w}^{\mathrm{H}} \mathbf{a}_{\theta_0,\phi_0,\gamma_0,\eta_0} = 1. \tag{38}$$

And the solution to (38) is commonly known as the MVDR beamformer:

$$\mathbf{w}_{\mathrm{opt}} = \frac{\mathbf{R}_{\mathbf{i+n}}^{-1} \mathbf{a}_{\theta_0,\phi_0,\gamma_0,\eta_0}}{\mathbf{a}_{\theta_0,\phi_0,\gamma_0,\eta_0}^{\mathrm{H}} \mathbf{R}_{\mathbf{i+n}}^{-1} \mathbf{a}_{\theta_0,\phi_0,\gamma_0,\eta_0}}. \tag{39}$$

The optimal output SINR can be expressed as $\sigma_0^2 \left|\mathbf{w}_{\mathrm{opt}}^{\mathrm{H}} \mathbf{a}_{\theta_0,\phi_0,\gamma_0,\eta_0}\right|^2 / \left(\mathbf{w}_{\mathrm{opt}}^{\mathrm{H}} \mathbf{R}_{\mathbf{i+n}} \mathbf{w}_{\mathrm{opt}}\right)$. However, accurately estimating the interference-plus-noise covariance matrix in practical scenarios poses challenges. To mitigate this, the estimation is often substituted by the sample matrix $\hat{\mathbf{R}}_{\mathbf{xx}} = \frac{1}{K}\sum_{k=1}^{K} \mathbf{x}(t_k)\mathbf{x}^{\mathrm{H}}(t_k)$, leading to the sample matrix inversion (SMI) beamformer [41], where K represents the number of snapshots. The SMI beamformer may introduce distortion, especially when the desired signal is present in training samples or due to model errors. Optimized robust beamformers like DLSMI and Eigenspace beamformers address these challenges but face issues such as factor selection and performance degradation at low input signal-to-noise ratios (SNRs). Further advancements include beamformers like the interference covariance matrix reconstruction beamformer and the interference covariance matrix sparse reconstruction beamformer. These are tailored for arrays without considering polarization diversity, potentially resulting in polarization mismatch.

To surpass these limitations, a beamformer operating in the joint spatial and polarimetric domain will be developed. This leverages the sparsity of source signals by estimating $\mathbf{a}_{\theta_0,\phi_0,\gamma_0,\eta_0}$ and reconstructing $\mathbf{R}_{\mathbf{i+n}}$. Once the estimated steering vector $\mathbf{a}_{\hat{\theta}_0,\hat{\phi}_0,\hat{\gamma}_0,\hat{\eta}_0}$ in obtained in (39), the remaining unknown is the interference-plus-noise covariance matrix $\mathbf{R}_{\mathbf{i+n}}$, represented as $\mathbf{R}_{\mathbf{i+n}} = \Sigma_{m=1}^{M}\sigma_m^2\mathbf{a}_{\theta_m,\phi_m,\gamma_m,\eta_m}\mathbf{a}^{\mathrm{H}}_{\theta_m,\phi_m,\gamma_m,\eta_m} + \sigma^2\mathbf{I}_{N_{\mathrm{m}}L}$. The reconstruction of $\mathbf{R}_{\mathbf{i+n}}$ involves both $\sigma^2\mathbf{I}_{N_{\mathrm{m}}L}$ and $\Sigma_{m=1}^{M}\sigma_m^2\mathbf{a}_{\theta_m,\phi_m,\gamma_m,\eta_m}\mathbf{a}^{\mathrm{H}}_{\theta_m,\phi_m,\gamma_m,\eta_m}$. Noise power estimation (denoted as $\hat{\sigma}^2$) can be derived from the $L_0N_{\mathrm{m}}N_{\mathrm{p}} - M - 1$ smaller eigenvalues of $\hat{\mathbf{R}}_{\mathbf{xx}}$. $\Sigma_{m=1}^{M}\sigma_m^2\mathbf{a}_{\theta_m,\phi_m,\gamma_m,\eta_m}\mathbf{a}^{\mathrm{H}}_{\theta_m,\phi_m,\gamma_m,\eta_m}$ can be determined using the power distribution $\mathbf{p} = \left[\sigma_0^2, \sigma_1^2, \cdots, \sigma_M^2\right]^{\mathrm{T}} \in \mathbb{R}^{(M+1)\times 1}$ [45]:

$$\min_{\mathbf{p}} \left\| \hat{\mathbf{R}}_{\mathbf{xx}} - \hat{\sigma}^2\mathbf{I}_{N_{\mathrm{m}}L} - \mathbf{C}\left(\hat{\boldsymbol{\theta}}, \hat{\boldsymbol{\phi}}, \hat{\boldsymbol{\gamma}}, \hat{\boldsymbol{\eta}}\right)\mathbf{P}\mathbf{C}^{\mathrm{H}}\left(\hat{\boldsymbol{\theta}}, \hat{\boldsymbol{\phi}}, \hat{\boldsymbol{\gamma}}, \hat{\boldsymbol{\eta}}\right)\right\|_{\mathrm{F}}^2 \tag{40}$$

in which $\mathbf{P} = \mathrm{diag}(\mathbf{p})$, and

$$\mathbf{C}\left(\hat{\boldsymbol{\theta}}, \hat{\boldsymbol{\phi}}, \hat{\boldsymbol{\gamma}}, \hat{\boldsymbol{\eta}}\right) = \left[\mathbf{a}_{\hat{\theta}_0,\hat{\phi}_0,\hat{\gamma}_0,\hat{\eta}_0}, \mathbf{a}_{\hat{\theta}_1,\hat{\phi}_1,\hat{\gamma}_1,\hat{\eta}_1}, \cdots, \mathbf{a}_{\hat{\theta}_M,\hat{\phi}_M,\hat{\gamma}_M,\hat{\eta}_M}\right] \tag{41}$$

incorporates the estimated steering vectors of the desired signal and interferences within the joint spatial and polarimetric domain, utilizing the estimated multi-dimensional parameters. The closed-form solution to (40) is provided by

$$\mathbf{p} = \left(\mathbf{B}^{\mathrm{H}}\mathbf{B}\right)^{-1}\mathbf{B}^{\mathrm{H}}\mathbf{r}, \tag{42}$$

where

$$\mathbf{B} = \left\{\mathrm{vec}\left[\mathbf{a}_{\hat{\theta}_0,\hat{\phi}_0,\hat{\gamma}_0,\hat{\eta}_0}\mathbf{a}^{\mathrm{H}}_{\hat{\theta}_0,\hat{\phi}_0,\hat{\gamma}_0,\hat{\eta}_0}\right], \mathrm{vec}\left[\mathbf{a}_{\hat{\theta}_1,\hat{\phi}_1,\hat{\gamma}_1,\hat{\eta}_1}\mathbf{a}^{\mathrm{H}}_{\hat{\theta}_1,\hat{\phi}_1,\hat{\gamma}_1,\hat{\eta}_1}\right], \cdots, \mathrm{vec}\left[\mathbf{a}_{\hat{\theta}_M,\hat{\phi}_M,\hat{\gamma}_M,\hat{\eta}_M}\mathbf{a}^{\mathrm{H}}_{\hat{\theta}_M,\hat{\phi}_M,\hat{\gamma}_M,\hat{\eta}_M}\right]\right\}, \tag{43}$$

and $\mathbf{r} = \mathrm{vec}\left(\hat{\mathbf{R}}_{\mathbf{xx}} - \hat{\sigma}^2\mathbf{I}_{N_{\mathrm{m}}L}\right)$. Therefore, the interference-plus-noise covariance matrix can be reconstructed as

$$\tilde{\mathbf{R}}_{\mathbf{i+n}} = \sum_{m=1}^{M}[\mathbf{p}]_{m+1}\mathbf{a}_{\hat{\theta}_m,\hat{\phi}_m,\hat{\gamma}_m,\hat{\eta}_m}\mathbf{a}^{\mathrm{H}}_{\hat{\theta}_m,\hat{\phi}_m,\hat{\gamma}_m,\hat{\eta}_m} + \hat{\sigma}^2\mathbf{I}_{N_{\mathrm{m}}L}. \tag{44}$$

The PSR beamformer is defined by substituting (44) and $\mathbf{a}_{\hat{\theta}_0,\hat{\phi}_0,\hat{\gamma}_0,\hat{\eta}_0}$ into (39), yielding:

$$\mathbf{w}_{\mathrm{P}} = \frac{\tilde{\mathbf{R}}_{\mathbf{i+n}}^{-1}\mathbf{a}_{\hat{\theta}_0,\hat{\phi}_0,\hat{\gamma}_0,\hat{\eta}_0}}{\mathbf{a}^{\mathrm{H}}_{\hat{\theta}_0,\hat{\phi}_0,\hat{\gamma}_0,\hat{\eta}_0}\tilde{\mathbf{R}}_{\mathbf{i+n}}^{-1}\mathbf{a}_{\hat{\theta}_0,\hat{\phi}_0,\hat{\gamma}_0,\hat{\eta}_0}}. \tag{45}$$

By utilizing the estimated multi-parameters detailed in Section 4, the interference-plus-noise covariance matrix can be reconstructed, leveraging the sparse characteristics of both the desired signal and interferences within the combined spatial and polarimetric domain.

It is crucial to emphasize that the PSR beamformer, supported by DSPR, operates under the assumption that the precise multi-dimensional parameters of the desired signal and interferences are unknown and require estimation. In contrast, non-blind beamformers, which operate with known parameters, inherently possess capabilities for suppressing main-lobe interference. This capability arises from their ability to process data simultaneously across both spatial and polarimetric domains, a fundamental aspect of the PSR beamformer's functionality.

6. Simulation results

In this section, six sets of simulations are carried out to analysis the mutual coupling effect of the considered sparse polarimetric array (i.e., CSPA in this chapter) and its associated collocated ones, validate the enhanced DoFs of the CSPA, investigate the performance of the DSPR supported PSR beamformer, and compare it with other competing methods. The competing methods include the SMI beamformer, DLSMI beamformer, Eigenspace beamformer, worst-case beamformer, interference covariance matrix sparse reconstruction beamformer (referred to as reconstruction beamformer for simplicity later on), and Integral-4D beamformer. All the competing beamformers are extended to the joint spatial and polarimetric domain, except for the Integral-4D beamformer originally designed in the joint spatial and polarimetric domain. It is important to note that the competing beamformers used for comparison employ the same CSPA, except for the reconstruction beamformer (originally proposed for 1D array primitively without considering polarization diversity) which uses an associated CSPA composed of unipolarized antennas. The diagonal loading factor is set as $10\sigma^2$ for the DLSMI beam-former, and $\varepsilon = 1$ is adopted for the worst-case beamformer. The desired signal, interferences, and the additive noise are modeled as complex circularly symmetric Gaussian zero-mean spatially and temporally white processes. Three simulation examples are considered: one with exactly known steering vectors, another with random DOA and polarization mismatch of the desired signal and interferences, and the third with mismatch caused by coherent local scattering. Each scenario is implemented with 500 Monte-Carlo trials. Since the Integral-4D beamformer requires precisely known parameters of both the desired signal and interferences, it is only included in the simulation example with exactly known steering vectors.

In the plots of output SINR against input SNR, the number of snapshots is set to 50. In the performance comparison of output SINR against the number of snapshots, the SNR is set at 15 dB. The interference-to-noise ratio (INR) is set at 20 dB (relatively high) in Subsections 6.4 to 6.6, while it is set as 20 and 5 dB (relative low) in Subsection 6.6 to compare the performance of the competing beamformers under different INRs. When the estimated parameters $\{\hat{\theta}_m\}_{m=0}^{M}$, $\{\hat{\phi}_m\}_{m=0}^{M}$, $\{\hat{\gamma}_m\}_{m=0}^{M}$, and $\{\hat{\eta}_m\}_{m=0}^{M}$ are out of the ranges $\{[\theta_m - 4.5°, \theta_m + 4.5°]\}_{m=0}^{M}$, $\{[\phi_m - 4.5°, \phi_m + 4.5°]\}_{m=0}^{M}$, $\{[\gamma_m - 4.5°, \gamma_m + 4.5°]\}_{m=0}^{M}$, and $\{[\eta_m - 4.5°, \eta_m + 4.5°]\}_{m=0}^{M}$, especially at low SNRs or with few snapshots, the 2D DOA and polarization parameters are presumed by drawing independently and randomly from the specified ranges using prior knowledge. In the considered CSPA, both d and d_x are $\lambda/2$.

Before comparing the performance of the introduced beamformer with other competing beamformers, the mutual coupling effect, beampattern analysis, and the 2D DOA and polarization estimation performance are studied.

In the mutual coupling effect analysis in Subsection 6.1, $|\dot{\mathbf{a}}_0^{\mathrm{H}}\mathbf{a}_0|/(\|\dot{\mathbf{a}}_0\|\|\mathbf{a}_0\|)$ is used to compare the cosine value of the angle between the presumed steering vector $\mathbf{a}_0$ and the realistic steering vector $\dot{\mathbf{a}}_0$ obtained from full-wave electromagnetic simulation. For simplicity and without loss of generality, $\gamma = 75^\circ$ and $\eta = -120^\circ$ are assumed for the impinging signal under consideration. In Subsection 6.3 (for a determined scenario) and Subsections 6.4 to 6.6, the parameters $\{\theta_0, \phi_0, \gamma_0, \eta_0\}$ of the desired signal are set as $\{8^\circ, 23^\circ, 83^\circ, -81^\circ\}$, while the parameters $\{\theta_m, \phi_m, \gamma_m, \eta_m\}_{m=1}^{2}$ of the two interferences are $\{32^\circ, 41^\circ, 23^\circ, 31^\circ\}$ and $\{81^\circ, 61^\circ, 66^\circ, -6^\circ\}$. Another case with four interferences is included in Subsection 6.5 to validate the effectiveness of the DSPR-supported PSR beamformer in different scenarios, where two additional interferences have parameters $\{\theta_m, \phi_m, \gamma_m, \eta_m\}_{m=3}^{4}$ being $\{44^\circ, 46^\circ, 44^\circ, 10^\circ\}$ and $\{61^\circ, 55^\circ, 16^\circ, -50^\circ\}$, respectively.

To verify the CSPA's insensitivity to mutual coupling, the generality of the DSPR method, and the (DSPR-supported) PSR beamformer, different simulation examples with varying array configurations and sizes (by adjusting N_{m} and N_{p}) are considered. The y-axis coordinates for the five elements in Sub-array 1 and Layer 1 are $\mathbf{v}_{1,1} = [0\ \ d\ \ 2d\ \ 5d\ \ 8d]^{\mathrm{T}}$, and only three elements ($\mathbf{v}_{1,1} = [0\ \ d\ \ 3d]^{\mathrm{T}}$) in Sub-array 1 and Layer 1 are used to verify the effectiveness for the underdetermined scenario in Subsections 6.2.

Four CSPA configurations are considered: Subsection 6.1 for Case 1) $N_{\mathrm{m}} = 2, N_{\mathrm{p}} = 3$, $\mathbf{J} = \mathrm{blkdiag}(\mathbf{I}_2, [0\ \ 1\ \ 0\ \ 0])$, $\mathbf{d}^{(l)}, l = 1, 2, \cdots, 15$ in the first layer are $\mathbf{d}^{(1)} = \cdots = \mathbf{d}^{(5)} = [1, 0, 0]^{\mathrm{T}}$, $\mathbf{d}^{(6)} = \cdots = \mathbf{d}^{(10)} = [0, 1, 0]^{\mathrm{T}}$, and $\mathbf{d}^{(11)} = \cdots = \mathbf{d}^{(15)} = [0, 0, 1]^{\mathrm{T}}$; Subsections 6.2 to 6.3 for Case 2) $N_{\mathrm{m}} = N_{\mathrm{p}} = 2$, $\mathbf{J} = [\mathbf{I}_2, \mathbf{O}_{2\times 4}]$, $\mathbf{d}^{(1)} = \cdots = \mathbf{d}^{(5)} = [1, 0]^{\mathrm{T}}$, $\mathbf{d}^{(6)} = \cdots = \mathbf{d}^{(10)} = [0, 1]^{\mathrm{T}}$; Subsection 6.4 for Case 3) $N_{\mathrm{m}} = 3, N_{\mathrm{p}} = 2$, $\mathbf{J} = [\mathbf{I}_2, \mathbf{O}_{2\times 4}]$, $\mathbf{d}^{(l)}, l = 1, 2, \cdots, 10$ in the first layer are the same as those in the Case 2); Subsections $6.5 \sim 6.6$ for Case 4) $N_{\mathrm{m}} = 2, N_{\mathrm{p}} = 3$, $\mathbf{J} = [\mathbf{I}_3, \mathbf{O}_3]$, $\mathbf{d}^{(l)}, l = 1, 2, \cdots, 15$ in the first layer are the same as those in Case 1).

6.1 Mutual coupling effect analysis

The mutual coupling effect analysis is conducted for the CSPA, the collocated sparse triad array, and the collocated sparse EMVA array. By migrating the antennas in Sub-array 2 and Sub-array 3 into Sub-array 1 for the CSPA and considering the y-axis antenna locations for the collocated sparse EMVA array, a comparison is made. The cosine value of the angle between the presumed steering vector and the realistic steering vector, influenced by mutual coupling through full-wave electromagnetic simulation, is given here. The simulation results depicted in **Figure 9** reveal that the cosine value remains consistently close to 1 for the CSPA. In contrast, the cosine values for the collocated sparse triad array and the collocated sparse EMVA array are consistently below 1. Specifically, the relatively lower cosine value observed for the collocated sparse EMVA array indicates that it exhibits the highest level of mutual coupling among the three arrays under comparison. On the other hand, due to its non-collocated structure achieved through a cascading manner, the CSPA demonstrates lower susceptibility to mutual coupling compared to the other two sparse arrays.

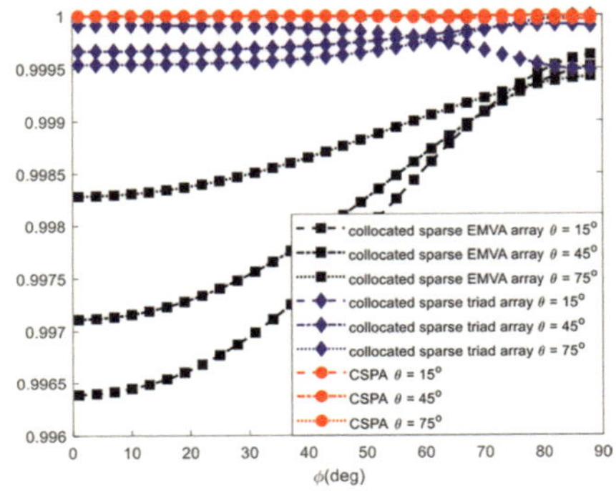

Figure 9.
Cosine value of the angle between the presumed steering vector and the realistic steering vector.

6.2 Closed-form auto-paired 2D DOA and polarization estimation analysis for both undetermined and determined scenarios

Initially, the underdetermined scenario involving 11 signals is considered. The 2D DOAs of these signals are $(11^\circ + 15^\circ r_1, 30^\circ + 10^\circ r_1)$, $r_1 = 0, 1, 2$, $(167^\circ - 15^\circ r_2, 13^\circ + 10^\circ r_2)$, $r_2 = 0, 1, 2, 3, 4$, $(59^\circ, 53^\circ)$, $(74^\circ, 63^\circ)$, and $(98^\circ, 71^\circ)$, along with their respective polarization parameters. Specifically, the first 7 signals share one set of polarization parameters $(45^\circ, 80^\circ)$, while the last 4 signals share one set of polarization parameters $(45^\circ, -80^\circ)$. In **Figure 10**, the 2D scatters for the closed-form auto-paired 2D DOA estimation are depicted. The analysis involves 200 Monte-Carlo trials, an SNR of 10 dB, and a large snapshot number of 500,000, which is necessary to accurately model the real covariance matrix for the underdetermined scenario. With 12 diversely polarized antennas utilized in this setup, traditional methods like the polarimetric element-space root MUSIC [35] can typically identify up to 10 signals, whereas up to 14 signals could be identified for the considered scenario as the number of virtual antennas for the associated CUPA would be increased to 16, showcasing enhanced DoFs.

In the determined scenario involving 3 signals, the 2D scatters for the closed-form auto-paired 2D DOA and polarization estimation are presented in **Figure 11(a)**. This analysis includes 50 trials, 500 snapshots, and a SNR of 20 dB (with all interferences

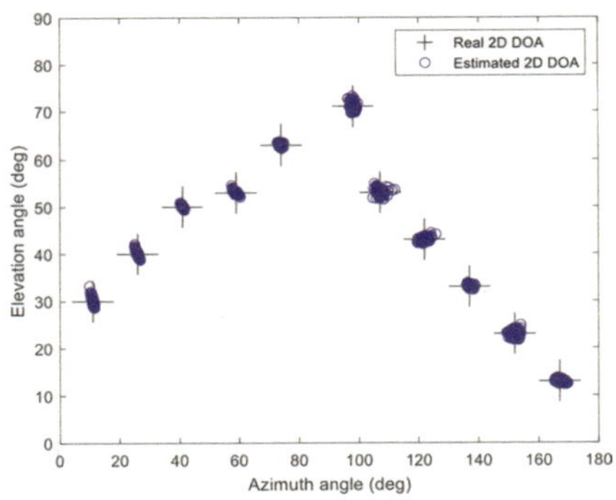

Figure 10.
2D DOA estimation scatter for the underdetermined scenario.

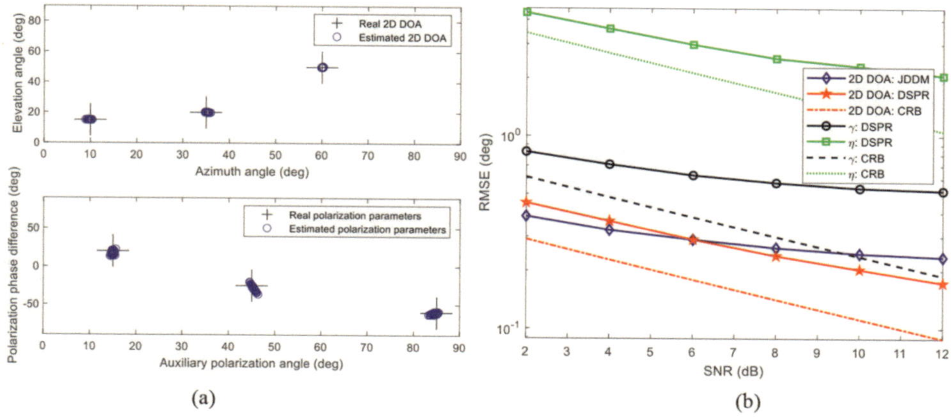

Figure 11.
2D DOA and polarization estimation performance for the determined scenario. (a) the scatters for the 2D DOA and polarization estimation; (b) RMSE versus the SNR.

and the desired signal having equal power). The results from the multi-parameter estimation scatters confirm the validity of the closed-form auto-paired property and the effectiveness of the DSPR method. **Figure 11(b)** displays statistical curves showing the root mean square error (RMSE) versus SNR with 800 snapshots. Additionally, a comparison with the joint diagonalization direction matrix (JDDM) method based on the reconstructed covariance matrix associated with the virtual CUPA is provided. The JDDM method exhibits subpar performance at high SNRs due to sensitivity in the reconstruction process. In contrast, the DSPR method maintains a small gap with the Cramér-Rao Bound derived from [53] across all SNRs, albeit with higher computational complexity.

Furthermore, the average computational time for each trial using the DSPR method is approximately 5.2 ms in MATLAB 2020a (on an Intel(R) Core(TM) i7-10875H 2.3 GHz CPU). In comparison, the conventional 2D spectral searching approach requires about 1.9 seconds per trial with a searching grid size of $\theta \sim 0 : 0.2^\circ : 90^\circ$ and $\phi \sim 0 : 0.2^\circ : 90^\circ$. This comparison highlights the significant computational advantage of the DSPR method.

6.3 Analysis on beampattern in joint spatial and polarimetric domain

The performance of the PSR beamformer is compared with that of the associated beamformer for the CSPA, which comprises uni-polarized antennas oriented with their dipoles parallel to the y-axis. In this comparison, the SNR of the desired signal and the INR of the two interferences were set at 30 dB, with 50 snapshots considered. The elevation and auxiliary polarization angle are set to the same value for ease of performance evaluation. The 2D DOA and polarization parameters of the desired signal and interferences are $(\theta_0, \phi_0, \gamma_0, \eta_0) = (43^\circ, 12^\circ, 45^\circ, 10^\circ)$, $(\theta_1, \phi_1, \gamma_1, \eta_1) = (43^\circ, 12^\circ, 45^\circ, 60^\circ)$, and $(\theta_2, \phi_2, \gamma_2, \eta_2) = (70^\circ, 12^\circ, 45^\circ, 40^\circ)$, indicating that the desired signal and the first interference share the same 2D DOA. In **Figure 12 (a)**, a valley is observed at a certain point $(\theta_2, \phi_2, \gamma_2, \eta_2)$ as expected, while no valley is present at $(\theta_1, \phi_1, \gamma_1, \eta_1)$ since the first interference and the desired signal share the same 2D DOA but different polarization parameters. **Figure 12(b)** demonstrates that

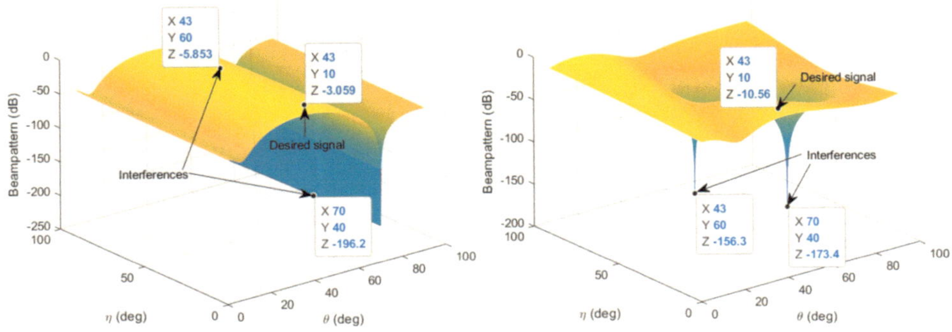

(a) (b)

Figure 12.
Beampattern for the compared arrays. (a) beampattern for the CSPA consisting of uni-polarized antennas; (b) beampattern for the CSPA consisting of diversely polarized antennas.

the PSR beamformer for the CSPA effectively suppresses the two interferences very well. In contrast, the associated beamformer for the CSPA, composed of uni-polarized antennas, lacks the capability to handle the polarization diversity of source signals, leading to inferior performance. This validates the superiority of the PSR beamformer for the CSPA over the associated beamformer for the CSPA with uni-polarized antennas.

6.4 Statistical beamforming performance for exactly known steering vector

In the scenario where the steering vectors of the desired signal and interferences are known, the performance of the PSR beamformer is evaluated against the optimal output SINR. A comparison is made with the reconstruction beamformer for the CSPA, which consists of uni-polarized antennas. From the analysis presented in **Figure 13**, it is evident that the PSR beamformer closely approaches the optimal output SINR across a wide range of input SNRs and snapshots. This is attributed to the accurate reconstruction of the interference-plus-noise covariance matrix in the joint

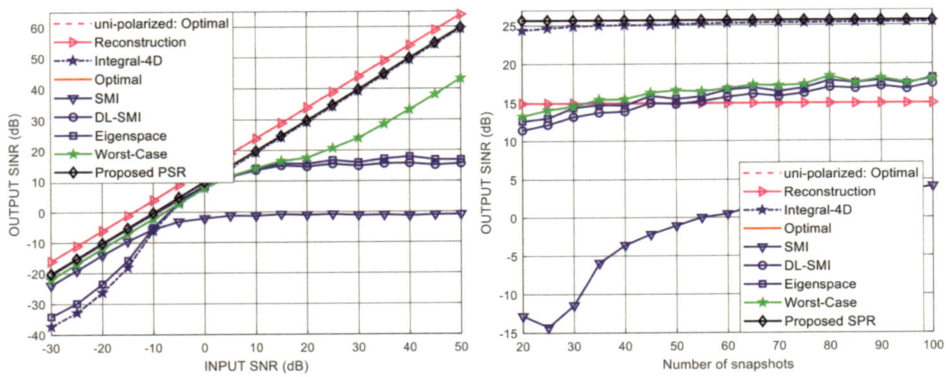

(a) (b)

Figure 13.
Exactly known desired signal steering vector. (a) Output SINR versus the input SNR; (b) Output SINR versus the number of snapshots.

spatial and polarimetric domain. The PSR beamformer demonstrates superior performance compared to other beamformers such as SMI, DLSMI, Eigenspace, and worst-case beamformers, particularly at high input SNRs and across all simulated snapshots where the desired signal may be attenuated in the comparison methods. The Integral-4D beamformer shows comparable performance to the PSR beamformer at high SNRs but deteriorates at low SNRs and with fewer snapshots due to suboptimal noise-subspace estimation. Real run-time comparisons reveal that the Integral-4D beamformer is significantly less efficient computationally, with a runtime of about 13.56 s compared to 0.59 ms for the PSR beamformer under the same simulation conditions. In the case of the CSPA with uni-polarized antennas, the performance of the reconstruction beamformer closely aligns with the associated optimal output SINR. However, there is a performance loss of approximately 10 dB for the CSPA with uni-polarized antennas due to polarization mismatch, highlighting the importance of considering polarization diversity in antenna array design for improved performance.

6.5 Statistical beamforming performance for unknown steering vector mismatch caused by random 2D DOA and polarization estimation error

This impact of the steering vector mismatch resulting from random errors in the desired signal and interference steering vectors will be considered. The 2D DOA and polarization parameters of both the desired signal and interferences follow $\mathcal{N}(\theta_m, 2^\circ)$, $\mathcal{N}(\phi_m, 2^\circ)$, $\mathcal{N}(\gamma_m, 2^\circ)$, and $\mathcal{N}(\eta_m, 2^\circ)$ for each trial, respectively, where $m = 0, 1, \cdots, 4$. In comparison with the findings in **Figure 13**, the performance degradation of various competing beamformers due to the steering vector mismatch is illustrated in **Figure 14**. The curves in **Figure 14** highlight how the competing beamformers experience performance deterioration because of the steering vector mismatch. Specifically, the performance curve of the PSR beamformer supported by the DSPR deviates from the optimal output SINR due to the non-ideal parameters in the multi-parameter estimation and reconstruction process. Furthermore, the performance of the DSPR-supported PSR beamformer is more significantly impacted in the presence of four interferences compared to two interferences, as the increased number of parameters with estimation errors leads to greater reconstruction errors. Despite this, the PSR beamformer's performance remains superior to other competing

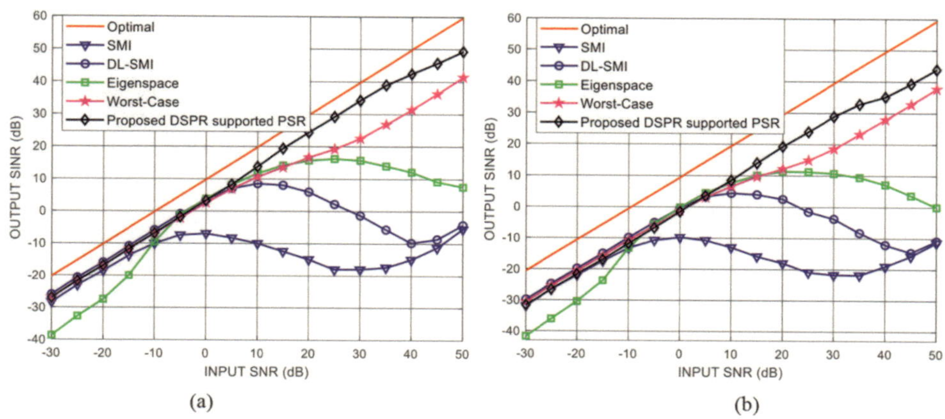

Figure 14.
Random DOA and polarization mismatch. (a) Output SINR versus the input SNR for two interferences; (b) Output SINR versus the number of snapshots for four interferences.

beamformers at high SNRs, and comparable to the DL-SMI and worst-case beamformers at low SNRs, showcasing the effectiveness of the reconstruction framework in the joint spatial and polarimetric domain.

6.6 Statistical beamforming performance for unknown steering vector mismatch caused by coherent local scattering

This example explores the impact of steering vector mismatch caused by coherent local scattering. The real steering vector comprises five components:

$$\bar{\mathbf{a}}_{\theta_0,\phi_0,\gamma_0,\eta_0} = \mathbf{a}_{\theta_0,\phi_0,\gamma_0,\eta_0} + \sum_{n=1}^{4} e^{j\delta_n}\mathbf{a}_{\theta_n,\phi_n,\gamma_n,\eta_n} \tag{46}$$

in which $\mathbf{a}_{\theta_0,\phi_0,\gamma_0,\eta_0}$ is associated with the direct path, and $\{\mathbf{a}_{\theta_n,\phi_n,\gamma_n,\eta_n}\}_{n=1}^4$ are corresponding to the four coherently scattered paths. The 2D DOA and polarization parameters ($\{\theta_n\}_{n=1}^4$, $\{\phi_n\}_{n=1}^4$, $\{\gamma_n\}_{n=1}^4$, and $\{\eta_n\}_{n=1}^4$) follow normal distributions with specific mean and variance values for each trial, i.e., $\mathcal{N}(\theta_0, 2^{\circ})$, $\mathcal{N}(\phi_0, 2^{\circ})$, $\mathcal{N}(\gamma_0, 2^{\circ})$, and $\mathcal{N}(\eta_0, 2^{\circ})$. $\{\delta_n\}_{n=1}^4$ represents the path phases that are independently and uniformly distributed in $[0, 2\pi]$ for each trial. Notably, the parameters $\{\theta_n, \phi_n, \gamma_n, \eta_n\}_{n=1}^4$ remain constant across snapshots but vary for each trial. In contrast to the findings in Subsections 6.4 and 6.5, **Figure 15** demonstrates a 7.2 dB increase in the optimal output SINR due to the multipath effect. The performance of SMI, DL-SMI, and Eigenspace beamformers deteriorates at high SNRs, attributed to the disturbed covariance matrix resulting from coherent local scattering.

The DSPR-supported PSR beamformer exhibits performance close to the optimal output SINR, particularly at low SNRs, for two key reasons. Firstly, at relatively high input SNRs, the multi-parameter estimation of the interferences can be achieved with high accuracy. Secondly, the coherent local scattering sources close to the desired source signal aid in the multi-parameter estimation of the desired signal, leading to precise noise power estimation. These factors contribute to the accurate estimation of the weighting vector in (45). However, when the input SNR exceeds the INR (20 dB

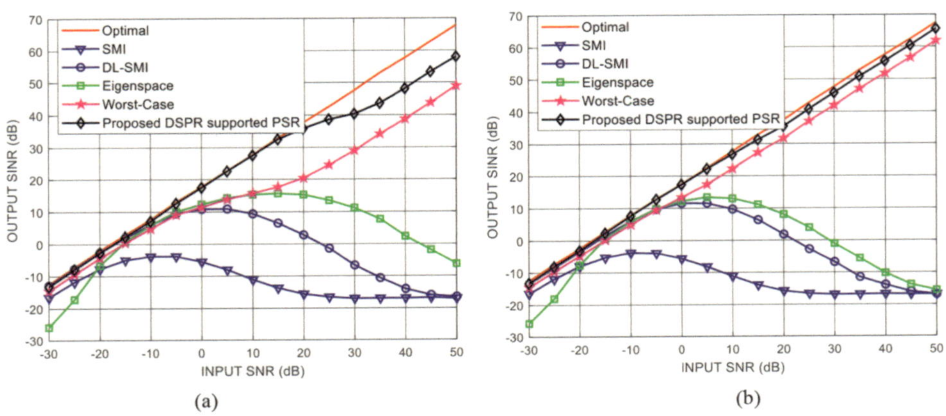

Figure 15.
Signal spatial signature mismatch caused by coherent local scattering. (a) Output SINR versus the input SNR when INR = 20 dB; (b) Output SINR versus the input SNR when INR = 5 dB.

in **Figure 15(a)**), the higher-power coherent local scattering sources may be misinterpreted as interferences, leading to a deterioration in the estimation of the interference-plus-noise covariance matrix. Consequently, the output SINR may not significantly increase with higher input SNRs. Similar performance degradation is observed when the input SNR surpasses 5 dB in **Figure 15(b)**, albeit less pronounced due to the relatively low INR.

7. Conclusions

The sparse polarimetric array has many advantages such as low mutual coupling and high DoFs, which can be used as future EL-MIMO-TAs to achieve functions such as integrated sensing and communication, high-precision perception, etc. This chapter systematically presents the design strategies (i.e., non-uniform, uniform, and hybrid non-uniform and uniform sparsity based on the sparsity of array element positions) of sparse polarimetric arrays, a closed-form multi-dimensional parameter estimation method based on the reconstructed covariance matrix via multi-dimensional data fitting, and a multi-dimensional beamformer in the spatial-polarimetric joint domain based on the sparse characteristic of source signals and provides simulation results to verify the advantages of sparse polarimetric arrays in the above aspects. It has been shown that the CSPA exhibits lower mutual coupling effects, enhanced DoFs by leveraging the proposed closed-form multi-parameter estimation approach, and improved beamforming performance when compared to current state-of-the-art beamformers. Such findings lay the foundation for practical applications in future MIMO systems.

Acknowledgements

This work was supported by the National Natural Science Foundation of China (grant numbers 62431021 and 62271414), Zhejiang Provincial Outstanding Youth Science Foundation (grant number LR23F010001), Zhejiang "Pioneer" and "Leading Goose" R&D Program (grant number 2024SDXHDX0006, 2024C03182), the Key Project of Westlake Institute for Optoelectronics (grant number 2023GD007), and Ningbo Science and Technology Bureau, "Science and Technology Yongjiang 2035" Key Technology Breakthrough Program (grant number 2024Z126).

Author details

Yaxing Yue[1,2]*, Guisheng Liao[1,2], Xin Yuan[3] and Zhiguo Shi[4]

1 Hangzhou Institute of Technology, Xidian University, Hangzhou, China

2 National Key Laboratory of Radar Signal Processing, Xidian University, Xi'an, China

3 School of Engineering, Westlake University, Hangzhou, China

4 College of Information Science and Electronic Engineering, Zhejiang University, Hangzhou, China

*Address all correspondence to: yueyaxing@xidian.edu.cn

References

[1] Wang X, Wan L, Huang M, Shen C, Zhang K. Polarization channel estimation for circular and non–circular signals in massive MIMO systems. IEEE Journal of Selected Topics in Signal Processing. 2019;**13**(5):1001-1016

[2] Jo O, Kim J, Yoon J, Choi D, Hong W. Exploitation of dual-polarization diversity for 5G millimeter-wave MIMO beamforming systems. IEEE Transactions on Antennas and Propagation. 2017;**65**(12):6646-6655

[3] Zhang J, Kim KJ, Glazunov AA, Wang Y, Ding L, Zhang J. Generalized polarization-space modulation. IEEE Transactions on Communications. 2020; **68**(1):258-273

[4] Yue Y, Xu Y, Liu Z. Two-dimensional direction-of-arrival estimation in monostatic MIMO radar. International Conference on Information Communication and Signal Processing. Shanghai, China. 2021:60-64

[5] Xia T. Joint diagonalization based DOD and DOA estimation for bistatic MIMO radar. Signal Processing. 2015; **108**:159-166

[6] Arora A, Tsinos CG, Rao BSMR, Chatzinotas S, Ottersten B. Hybrid transceivers design for large-scale antenna arrays using majorization-minimization algorithms. IEEE Transactions on Signal Processing. 2020; **68**:701-714

[7] Yue Y, Xu Y, Zhuang J, Huang Y, Zhao K, Liu Z. Mutual coupling self-calibration for parameter estimation with vector antennas. In: IEEE International Conference on Signal, Information and Data Processing. Chongqing, China. 2019. pp. 1-5

[8] Shen Q, Liu W, Cui W, Wu S, Pal P. Simplified and enhanced multiple level nested arrays exploiting high order difference co-arrays. IEEE Transactions on Signal Processing. 2019;**67**(13):3502-3515

[9] Yang M, Ding J, Chen B, et al. Coprime L-shaped array connected by a triangular spatially-spread electromagnetic vector-sensor for two-dimensional direction of arrival estimation. IET Radar Sonar and Navigation. 2019;**13**(10):1609-1615

[10] Yue Y, Xu Y, Liu Z. Manifold separation and polarimetric element space based parameter estimation for polarimetric monostatic MIMO radar. In: CIE Int. Conf. Radar, Haikou, Hainan, China. 2021. pp. 573-577

[11] Yue Y, Wang Y, Xing F, Shi Z, Liao G. Polynomial rooting-based parameter estimation for polarimetric monostatic MIMO radar. Signal Processing. 2023;**2023**:109172

[12] Gong X, Jiang J, Li H, et al. Spatially spread dipole/loop quint for vector-cross-product-based direction finding and polarisation estimation. IET Signal Processing. 2018;**12**(5):636-642

[13] Zhu D, Chen B, Yang M, Dong M. New algorithm for joint estimation of DOA and polarisation based on electrically large and complete electromagnetic vector sensor. IET Radar, Sonar and Navigation. 2020; **14**(9):1279-1286

[14] Sun Y, Xie J, Han C, Wang L, Tao M. Array element selection strategies for interference suppression in reconfigurable tripole antenna array systems. IEEE Transactions on Vehicular Technology. 2023;**72**(1):557-572

[15] Yue Y, Li T, Zhou C, Yuan X, Shi Z. Research progress and prospect of sparse

diversely polarized array design. Journal of Radars. 2023;**12**(2):312-331

[16] Zhou C, Gu Y, He S, Shi Z. A robust and efficient algorithm for coprime array adaptive beamforming. IEEE Transactions on Vehicular Technology. 2018;**67**(2):1099-1112

[17] Amin MG, Wang X, Zhang YD, Ahmad F, Aboutanios E. Sparse arrays and sampling for interference mitigation and DOA estimation in GNSS. Proceedings of the IEEE. 2016;**104**(6): 1302-1317

[18] Yue Y, Xu Y, Liu Z. Closed-form two-dimensional DOA and polarization estimation of coexisted circular and noncircular signals. In: CIE Radar Conference, Haikou, China. 2021. pp. 1556-1560

[19] Shi Z, Zhou C, Gu Y, Goodman NA, Qu F. Source estimation using coprime array: A sparse reconstruction perspective. IEEE Sensors Journal. 2017; **17**(3):755-765

[20] Zhou C, Gu Y, Fan X, Shi Z, Mao G, Zhang YD. Direction-of-arrival estimation for coprime array via virtual array interpolation. IEEE Transactions on Signal Processing. 2018;**66**(22): 5956-5971

[21] Pal P, Vaidyanathan PP. Nested arrays: A novel approach to array processing with enhanced degrees of freedom. IEEE Transactions on Signal Processing. 2010;**58**(8): 4167-4181

[22] Liu C-L, Vaidyanathan PP. Super nested arrays: Linear sparse arrays with reduced mutual coupling-part I: Fundamentals. IEEE Transactions on Signal Processing. 2016;**64**(15): 3997-4012

[23] Liu J, Zhang Y, Lu Y, Ren S, Cao S. Augmented nested arrays with enhanced DOF and reduced mutual coupling. IEEE Transactions on Signal Processing. 2017; **65**(21):5549-5563

[24] Si W, Zeng F, Qu Z, et al. Two-dimensional DOA estimation via a novel sparse array consisting of coprime and nested subarrays. IEEE Communications Letters. 2020;**24**(6):1266-1270

[25] Yang Y, Hou Y, Mao X, et al. Stokes parameters and DOA estimation for nested polarization sensitive array in unknown nonuniform noise environment. Signal Processing. 2020; **175**:1-10

[26] Han K, Nehorai A. Nested vector-sensor array processing via tensor modeling. IEEE Transactions on Signal Processing. 2014;**62**(10):2542-2553

[27] Zoltowski M, Wong KT. ESPRIT-based 2-D direction finding with a sparse uniform array of electromagnetic vector sensors. IEEE Transactions on Signal Processing. 2000;**48**(8):2195-2204

[28] Wong KT, Yuan X. Vector cross-product direction-finding with an electromagnetic vector-sensor of six orthogonally oriented but spatially noncollocating dipoles/loops. IEEE Transactions on Signal Processing. 2011; **59**(1):160-171

[29] Zheng G. Two-dimensional DOA estimation for polarization sensitive array consisted of spatially spread crossed-dipole. IEEE Sensors Journal. 2018;**18**(12):5014-5023

[30] Yue Y, Zhang Z, Zhou C, Xing F, Shi Z. Closed-form two-dimensional DOA and polarization joint estimation using non-collocated sparse COLD array. In: Proc. IEEE Sensor Array and Multichannel Signal Processing

Workshop, Trondheim, Norway. 2022. pp. 16-20

[31] Qin S, Zhang YD, Amin MG. Improved two-dimensional DOA estimation using parallel coprime arrays. Signal Processing. 2020;**172**:107428

[32] He J, Li L, Shu T. 2-D direction finding using parallel nested arrays with full co-array aperture extension. Signal Processing. 2021;**178**:107795

[33] Yue Y, Zhou C, Xing F, Choo K-KR, Shi Z. Adaptive beamforming for cascaded sparse diversely polarized planar array. IEEE Transactions on Vehicular Technology. 2023;**72**(12): 15648-15664

[34] Yue Y, Zhang Z, Zhou C, Wu Y, Xing F, Shi Z. Closed-form robust adaptive beamforming using sparse diversely polarized antenna array. In: IEEE International Symposium on Personal, Indoor and Mobile Radio Communications, Toronto, ON, Canada. 2023. pp. 1-6

[35] Yue Y, Xu Y, Liu Z, Shen L. Parameter estimation of coexisted circular and strictly noncircular sources using diversely polarized antennas. IEEE Communications Letters. 2018;**22**(9): 1822-1825

[36] Yue Y, Xu Y, Shen L, Liu Z. Parameter estimation of coexisted circular and strictly non-circular signals. Electronics Letters. 2017;**53**(13):864-866

[37] Yue Y, Xu Y, Liu Z. Root high-order cumulant MUSIC. Digital Signal Processing. 2022;**122**:103328

[38] Zhang L, Huang L, Li B, Huang M, Yin J, Bao W. Fast-moving jamming suppression for UAV navigation: A minimum dispersion distortionless response beamforming approach. IEEE Transactions on Vehicular Technology. 2019;**68**(8):7815-7827

[39] Yue Y, Zhang Z, Shi Z. Generalized widely linear robust adaptive beamforming: A sparse reconstruction perspective. IEEE Transactions on Aerospace and Electronic Systems. 2024. DOI: 10.1109/TAES.2024.3397240

[40] Liu W, Weiss S. Wideband Beamforming: Concepts and Techniques. U.K., Chichester: Wiley; 2010

[41] Liu J, Liu W, Liu H, Chen B, Xia X-G, Dai F. Average SINR calculation of a persymmetric sample matrix inversion beamformer. IEEE Transactions on Signal Processing. 2016;**64**(8):2135-2145

[42] Shaw A, Smith J, Hassanien A. MVDR beamformer design by imposing unit circle roots constraints for uniform linear arrays. IEEE Transactions on Signal Processing. 2021;**69**:6116-6130

[43] Gu Y, Leshem A. Robust adaptive beamforming based on interference covariance matrix reconstruction and steering vector estimation. IEEE Transactions on Signal Processing. 2012; **60**(7):3881-3885

[44] Chen P, Yang Y, Wang Y, Ma Y, Yang L. Robust covariance matrix reconstruction algorithm for time-domain wideband adaptive beamforming. IEEE Transactions on Vehicular Technology. 2019;**68**(2): 1405-1416

[45] Gu Y, Goodman NA, Hong S, Li Y. Robust adaptive beamforming based on interference covariance matrix sparse reconstruction. Signal Processing. 2014; **96**(Part B):375-381

[46] Zheng Z, Zheng Y, Wang W-Q, Zhang H. Covariance matrix reconstruction with interference steering

vector and power estimation for robust adaptive beamforming. IEEE Transactions on Vehicular Technology. 2018;**67**(9):8495-8503

[47] Liu C, Fang Z, Xiang S. Joint polarization and space domain adaptive beamforming for dual polarized conformal array. In: Proc. Int. Conf. Syst. Inform. 2019. pp. 1126-1130

[48] Dai H, Wang X, Li Y, Liu Y, Xiao S. Main-lobe jamming suppression method of using spatial polarization characteristics of antennas. IEEE Transactions on Aerospace and Electronic Systems. 2012;**48**(3): 2167-2179

[49] Yang X, Zhang Z, Zeng T, Long T, Sarkar TK. Mainlobe interference suppression based on eigen-projection processing and covariance matrix reconstruction. IEEE Antennas and Wireless Propagation Letters. 2014;**13**: 1369-1372

[50] Lan L, Xu J, Liao G, Zhang Y, Fioranelli F, So HC. Suppression of mainbeam deceptive jammer with FDA-MIMO radar. IEEE Transactions on Vehicular Technology. 2020;**69**(10): 11584-11598

[51] Ge M, Cui G, Yu X, Kong L. Mainlobe jamming suppression with polarimetric multi-channel radar via independent component analysis. Digital Signal Processing. 2020;**106**:102806

[52] Aubry A, Carotenuto V, De Maio A. A new optimality property of the capon estimator. IEEE Signal Processing Letters. 2017;**24**(11):1706-1708

[53] Wang M, Nehorai A. Coarrays, MUSIC, and the Cramér-Rao bound. IEEE Transactions on Signal Processing. 2017;**65**(4):933-946